A PLUM

THE PHYSIC

D0283090

DIANDRA LESLIE-PELECKY is professor of physics at the University of Texas at Dallas. Her research is funded by the Department of Energy, the National Science Foundation, and the National Institutes of Health. She lives in Dallas.

Praise for *The Physics of NASCAR*©

"The first part of the book deals with materials, and looks at how combustion, power, and aerodynamics work together to maximize speed. But it's the driver and his crew who win the race, and Leslie-Pelecky gets plenty of time with the men behind the machines, joining Ray Evernham's crew to watch him race, and taking a turn behind the wheel herself. Along the way, the nanotech specialist becomes an unlikely racing fan; this fun physics primer should give any NASCAR aficionado a similar appreciation for science."

—Publishers Weekly

"What changes do we make during the race? Why are we making them? *The Physics of NASCAR* helps answer all of these questions. Reading it is a bit like being around a team working on winning. Read on and learn about the science of our sport, and the magic will come." —Ray Evernham, winner of three NASCAR Cup championships as crew chief with Jeff Gordon

"Reading this book isn't like watching a NASCAR race—it's like being behind the wheel. Or under the hood, as power flows from the pistons to the track. Physics and racing are united by a common respect for how the real world works, and Diandra Leslie-Pelecky does a great job at illuminating it in entertaining style."

—Sean Carroll, theoretical physicist, California Institute of Technology

THE PHYSICS OF NASCAR®

The Science Behind the Speed

Diandra Leslie-Pelecky

Foreword by Ray Evernham

A PLUME BOOK

PLUME
Published by the Penguin Group
Penguin Group (USA) Inc., 375 Hudson Street, New York, New York 10014, U.S.A. • Penguin Group (Canada), 90 Eglinton Avenue East, Suite 700, Toronto, Ontario, Canada M4P 2Y3 (a division of Pearson Penguin Canada Inc.) • Penguin Books Ltd., 80 Strand, London WC2R 0RL, England • Penguin Ireland, 25 St. Stephen's Green, Dublin 2, Ireland (a division of Penguin Books Ltd.) • Penguin Group (Australia), 250 Camberwell Road, Camberwell, Victoria 3124, Australia (a division of Pearson Australia Group Pty. Ltd.) • Penguin Books India Pvt. Ltd., 11 Community Centre, Panchsheel Park, New Delhi – 110 017, India • Penguin Group (NZ), 67 Apollo Drive, Rosedale, North Shore 0632, New Zealand (a division of Pearson New Zealand Ltd.) • Penguin Books (South Africa) (Pty.) Ltd., 24 Sturdee Avenue, Rosebank, Johannesburg 2196, South Africa

Penguin Books Ltd., Registered Offices: 80 Strand, London WC2R 0RL, England

Published by Plume, a member of Penguin Group (USA) Inc. Previously published in a Dutton edition.

First Plume Printing, February 2009

10 9 8 7 6 5 4 3 2 1

Image on page 14 adapted from NASCAR; all other images courtesy of the author.

Ⓟ REGISTERED TRADEMARK—MARCA REGISTRADA

THE LIBRARY OF CONGRESS HAS CATALOGUED THE DUTTON EDITION AS FOLLOWS:

Leslie-Pelecky, Diandra L.
The physics of NASCAR: how to make steel + gas + rubber - speed / Diandra Leslie-Pelecky.
    p.   cm.
Includes bibliographical references and index.
ISBN 978-0-525-95053-0 (hc.)
ISBN 978-0-452-29022-8 (pbk.)
1. Stock cars (Automobiles)—Design and construction.   2. Stock cars (Automobiles)—Performance.   3. Stock cars (Automobiles)—Equipment and supplies.   4. Stock cars (Automobiles)—Dynamics.   5. Stock car racing.   6. NASCAR (Association).   I. Title.
TL236.28.L47 2008
796.7201'53—dc22   2007046081

Printed in the United States of America
Original hardcover design by Level C

PUBLISHER'S NOTE
While the author has made every effort to provide accurate telephone numbers and Internet addresses at the time of publication, neither the publisher nor the author assumes any responsibility for errors, or for changes that occur after publication. Further, publisher does not have any control over and does not assume any responsibility for author or third-party websites or their content.

The scanning, uploading, and distribution of this book via the Internet or via any other means without the permission of the publisher is illegal and punishable by law. Please purchase only authorized electronic editions, and do not participate in or encourage electronic piracy of copyrighted materials. Your support of the author's rights is appreciated.

BOOKS ARE AVAILABLE AT QUANTITY DISCOUNTS WHEN USED TO PROMOTE PRODUCTS OR SERVICES. FOR INFORMATION PLEASE WRITE TO PREMIUM MARKETING DIVISION, PENGUIN GROUP (USA) INC., 375 HUDSON STREET, NEW YORK, NEW YORK 10014.

To all the men and women who
work in the garage and at the race shops

Contents

Foreword

The purpose of racing is to win; you win by beating the competition to the finish line. There are several factors that put you in a position to win; some say it's luck, but honestly there's more to it than that! Key factors dictate your success—and success ultimately comes from the science behind these factors.

For a race team to win, performance relies upon the driver, the team, and the vehicle. The performance of a race car is a function of:

- Engine power
- Aerodynamics
- Tire grip capability

All of this and a crystal ball will put you in a position to win. Science and the application of its principals *is your crystal ball* . . . that's racing, it's that simple and that complex.

As a driver, I knew how the car felt to me, what it felt like to go fast. It was my job to communicate this feeling to the crew chief so he could adjust the car to keep that feeling throughout a race. The feeling was intuitive; the challenge was in the communication. How do you communicate what you feel and have the crew chief translate this into a mechanical change or adjustment?

When I was a crew chief with Jeff Gordon, we earned 3 championships, 47 wins, and 116 top-five finishes. For us it was part magic and part science. The dynamic of the team, how to work together to achieve a common goal, was certainly a big part of the success and felt like magic; the application of information on how we built the cars and how we raced the cars, decisions made during the race, that was science.

As a team owner, I look at my collective experience as a driver, mechanic, and crew chief, and identifyed a common denominator of success on the race track: It is the systematic application of information—the science; and the team working in synchronicity—the magic.

Racing is a thrill, in its simplest form—beating everyone to the finish line—to the most complex of "*How* did you beat everyone to the finish line?" "*What* changes do we make during the race?" and "*Why* are we making them?" *The Physics of NASCAR* helps answer all of these questions. Read on and learn about the science (and magic) of our sport.

Ray Evernham

Preface

The Physics of What . . . ?

Have you ever found yourself in the middle of something really exciting and realized that, instead of paying attention, you were thinking about how you got there?

It happened to me on a very cold February evening in Atlanta. To my left, Jeff Gordon was climbing into his No. 24 DuPont Monte Carlo. A few feet away to my right, Elliott Sadler, the driver of the car I had followed around Atlanta Motor Speedway for the last ten and a half hours, was being interviewed. Elliott's crew—a great group of guys who had spent the last ten and a half hours trying to work while I asked a lot of questions—debated tire pressures over the radio. Elliott climbed into his No. 19 Dodge Dealers/United Auto Workers Dodge Charger and fired up the engine. Sitting on the cold pit wall with the crew waiting to see how he would do on the track, I found myself wondering how I—a physics professor— ended up writing a book about NASCAR.

Absentmindedly flipping through channels the Sunday before Memorial Day 2005, I happened to see a group of six cars rounding the corner of a track. Before I could flip to the next channel, chaos ensued. The back of one car wiggled slightly, and then—WHAM! The car smacked against the outside wall, then careened back down across the track, taking out the other five cars. Brakes screeched, cars spun through the grass, and a giant cloud of smoke rose over the infield.

People like me go into science because we like understanding things—or maybe it's better to say that it bothers us when we don't understand things. I get cranky and frustrated when I can't figure things out. Why would one car, for seemingly no reason, all of a sudden hit the wall?

The lengthy cleanup provided plenty of opportunities to view the accident on replays, which allowed me to dismiss a number of explanations. The car that crashed wasn't going any faster than the other cars. There were no flat tires, no engine failures, and no contact from any of the other cars. As I watched the replays—and then the rest of the race—I became more and more curious.

How do you build an engine that has three times the horsepower of a standard car engine and can run at 9,000 rpm for three hours without blowing up? How do you protect a driver from the 1,800°F flames of a gasoline fire? How do you construct a car so that the driver not only survives a 190-mph crash, but remembers to plug all of his sponsors before being treated and released at the infield care center?

The people who work at and with NASCAR were generous in helping me satisfy my curiosity. Fabricators showed me how a pile of sheet metal and tubing becomes a car, and engine builders helped me appreciate why a NASCAR engine—even though it works on the same basic principles—is more than just a larger version of your car's engine. I learned about wind tunnels from aerodynamicists and seven-post rigs from vehicle dynamicists. I was treated to an inside look at how race teams use our constantly evolving understanding of math and science to battle the intractable laws of nature—and the slightly less intractable laws of NASCAR. Piloting a race car at 150 mph at Texas Motor Speedway showed me that driving a race car is much different than just driving your own car really fast—and it was really fun.

Today's NASCAR is a complex, technical sport that works at the limits of what we understand about aerodynamics, structural engineering, and even human physiology. Despite its complexity, the foundation of NASCAR science consists of a few basic principles that every NASCAR fan can appreciate. Crew members—even if they can't recite Newton's Laws of Motion—have used these principles since the sport began to make cars faster and safer. Although they might not use the same words I use, NASCAR drivers quickly develop an intuitive understanding of the principles of aerodynamics and kinematics—or they crash a lot.

I'm sure you've sat in front of your television, or in the grandstands, and wondered why your team isn't running well. This book will help you understand why your driver's crew chief is going up on tire pressure, adding wedge, lowering the trackbar, or taking two tires instead of four. In addition to enhancing your understanding of NASCAR, the same basic principles that apply to Dale Earnhardt, Jr. going 200 mph at Daytona also apply to you as you're driving around your neighborhood—hopefully, at a significantly lower speed.

When I started this project, I didn't fully appreciate how much science there is in NASCAR, or how quickly NASCAR changes. Just during the time I spent writing this book, the NASCAR NEXTEL Cup Series changed to the NASCAR Sprint Cup Series and the Car of Tomorrow became just "the car." My website (www.stockcarscience.com) contains more detailed explanations, references for additional reading, and updates.

The greatest reward from writing this book wasn't learning the technical marvels of NASCAR. It was the privilege of meeting an exceptional group of people who are as passionate about racing as I am about science. I hope you'll value the technical content of this book, but also that you'll also enjoy meeting the people who use math and science in the race shop and at the track to accomplish the ultimate NASCAR goal: winning races.

THE PHYSICS OF NASCAR®

One

Bones of Contention

A Boeing 757 touches down at about 170 mph, but it didn't seem particularly fast watching from my window seat as I landed at the Charlotte airport. As a physics professor, I calculate speeds all the time in my research and in classes. Speed is how far you've gone divided by how long it took you to go that far. I teach my students that it's important to develop a feel for what the numbers you calculate mean in terms of things you know. At 170 mph, you travel the length of a football field in a little more than a second. It's also about the speed at which a NASCAR race car enters turn three at Las Vegas Motor Speedway.

If you really want to understand speed, North Carolina is the place to start. More than 90 percent of the NASCAR Sprint Cup Series teams have shops within a hundred miles of Charlotte. A number of those shops are in Concord, which is about twenty miles north of the Charlotte airport on I-85. I took the Concord Mills exit. Looming off to my right was Lowe's Motor Speedway, but I headed left, past the giant outlet mall and toward the Roush Fenway Racing shop. Jack Roush, often called the "Cat in the Hat" for the ever-present wide-brimmed Panama headgear he favors, once taught math and physics at Monroe Community College in Michigan. Today, he is the driving force behind Roush Fenway Racing, one of the largest and most successful NASCAR teams.

At the end of Roush Place is a driveway that encircles a thirty-foot-wide cylindrical fountain atop which ROUSH is spelled out in foot-high, silver block letters. Water flows underneath the letters and down the polished black granite sides. Directly ahead are the corporate headquarters and museum, and just behind are the buildings in which cars for Roush's five NASCAR Sprint Cup teams are built and repaired.

The parking lot was busier than usual because it was a few days before the 2006 October race at Lowe's Motor Speedway. Displays and activities were set up outside for visiting fans. Near the main building, a family of four looked over a pile of crumpled yellow-and-black pieces of sheet metal guarded by a life-size cardboard cutout of driver Matt Kenseth. This is recycling, NASCAR style: Fans eagerly purchase car parts demolished by their favorite driver.

The museum displays trace the history of Roush Racing, which became Roush Fenway Racing at the start of the 2007 season. One roped-off exhibit contains the remains of the No. 60 ARCA RE/MAX car after a race at Daytona. A video shows how the car got to its current state. I realized only after I watched driver Todd Kluever walk away from the accident that I had been holding my breath while the car literally disintegrated as it barrel-rolled through the infield. The sheet metal—or what was left of it—was peeled away, revealing the tube frame and reminding me why I was there.

An elevator took me to the second floor, where the reception area looked like any corporate waiting room: brightly lit, with a high ceiling and lots of light-colored wood and frosted glass. The thing that tells you you're in a race shop is the receptionist on the phone, helping fans find the shop. This particular attempt was a little more confounding than usual because the fan had absolutely no idea where he was.

Jamie Rodway, Roush's director of licensing, was my guide. Jamie is tall and smiling, with dark hair and lots of enthusiasm. "The great thing," Jamie said as we drove over to one of the fabrication buildings, "is that I get to cheer for my job every weekend. Not many people get to do that." This sentiment was echoed at every shop I visited.

The fabrication-building lobby, which is open to the public, features twenty-foot-high windows that reveal a cavernous but brightly lit shop. Rows of cars in various stages of construction or repair sit in front of the window on an impossibly clean, gray-painted concrete floor.

We passed through a door to the left of the viewing window. The door and windows insulate the lobby from the shop noise. I occasionally had to lean in to hear Jamie over the periodic pinging of hammers on metal, the popping of rivets, and the occasional growl of a disc cutter slicing through a length of steel tubing. I expected noise: The surprise was the smell. Race shops have a characteristic aroma, which I learned later is a mixture of brake cleaner, which is used as an all-purpose cleaning fluid, and gear oil. You don't notice the smell after a few minutes, but it is striking when you first walk into a shop.

The shop floor at Roush Fenway Racing is littered with body parts—or, to be more accurate—chassis parts. The chassis is a carefully constructed arrangement of tubing that serves as the race car's skeleton. The chassis supports the car's body and protects the driver, just as your bones support the rest of your body and protect your organs.

Although NASCAR's full name is the National Association for Stock Car Auto Racing, little about these cars is "stock." NASCAR started in 1948 using regular passenger cars, but the Impala your mom drove to the grocery store wasn't really designed to run at high speed on rutted dirt tracks. NASCAR allowed drivers to modify their cars to make them stronger and—as cars got faster—safer.

The first modifications involved nothing more than replacing some factory parts with sturdier heavy-duty parts. Roll cages—tubular structures that surround the driver—were introduced in the 1960s. Eventually, production cars got smaller and switched to unibody frames, and at some point, it became easier to build a race car that looked like a stock car than to convert a stock car into a race car.

There was absolutely nothing stock about the row of cars in front of me. Each car is a handmade original with its own unique quirks and characteristics. When drivers talk about bringing a specific car to a track, they really mean the chassis. Sheet metal gets put on and cut off, and engines, springs, and shocks are installed and removed, but the chassis really *is* the car. Matt Kenseth drove Roush Chassis No. 10, which was built in 1999 during his rookie year, for the last time at the fall 2006 Bristol race. He might have driven it again if NASCAR hadn't switched car designs.

Racks and racks of steel tubing and sheet metal line the walls in this part of Roush's fabrication, or "fab," shop. NASCAR mandates the sizes (diameters and wall thicknesses) of every piece of tubing in the chassis, as well as the type of metal from which the tubing is made.

Metals define civilizations. The "iron age" of a culture is the point at which iron production is the most sophisticated form of metalworking. Iron is an element, which means that a piece of pure iron contains iron—and only iron—atoms. Scientists use abbreviations to refer to elements because they let us express our ideas faster (and we tend to be a slightly impatient lot). Iron, for example, is represented by "Fe"; it's "Fe" and not "Ir" because the abbreviation comes from the Latin *ferrum*. ("Fe" isn't much shorter than "iron," but "Mo" is a lot shorter than "molybdenum.") "Ferrari" comes from the same word—it's an Italian occupational surname that means "metalworker," similar to the English "Smith."

Iron is usually found mixed with other elements. We believe iron compounds were used as early as 4000 B.C. in metal spear tips and ornaments. That metal came from meteorites, which we know because items from those time periods contain nickel, which isn't present in iron mined from the Earth.

Iron in the Earth is most often found in a compound with oxygen. Again, we use abbreviations to indicate what atoms and how many of them there are in a compound. Fe_3O_4, for example, tells you that there are four oxygen (O) atoms for every three iron (Fe) atoms. Fe_2O_3 is the iron-oxygen compound that gives the planet Mars its distinctive red color.

The problem with iron oxides (a metal combined with oxygen is called an oxide) is that they tend to be weak and crumbly. After all, Fe_2O_3 is just a fancy expression for rust. Early on, people figured out that you could get rid of most of the oxygen by heating and cooling iron oxide in a pile of charcoal. As with much of science, we figured out *how* to do this before we understood *why* it worked.

Charcoal is mostly carbon. When iron oxides are heated in charcoal, a chemical reaction between carbon and oxygen removes oxygen from the iron oxide. Iron has a very high melting temperature (2,800°F), so the iron never gets hot enough to actually liquify. Removing the oxygen leaves a porous metallic glob that is mostly iron and, depending on where the iron oxide came from, possibly very small amounts of other elements. Forming this material into tools, weapons, or ornaments required brute force. Early blacksmiths had to repeatedly heat and pound a piece of metal into a sword or a horseshoe.

There are no open hearths or anvils in the Roush Fenway shop, but there are a lot of different metalworking tools. Shears and snips are used to cut sheet metal, and a large brake can turn a flat piece of metal into a box in just a few steps. You may have bent quarter-inch copper tubing for a plumbing project: Across the shop, a Roush Fenway fabricator is using a pneumatic tube bender to shape a one-and-a-half-inch-diameter steel tube, which he does with measured precision.

The metal used to build a race car has to be strong, but not so strong that you can't bend or otherwise coerce it into the shapes you need to make the car. We have to balance ductility (how easy it is to shape a material without breaking it) and strength (how well a material retains its shape when pushed or pulled). These two characteristics are determined by how the atoms in the material are organized.

The atoms in a metal are arranged in a regular pattern: cubes, or maybe parallelograms, depending on the type of metal. These regular arrangements of atoms are called crystals. The atoms are connected to each other by chemical bonds, which (in solids) behave very much like springs, as shown in the drawing below. When you press on a solid, you compress the springs between the atoms, just like you compress the springs in your mattress when you lie down.

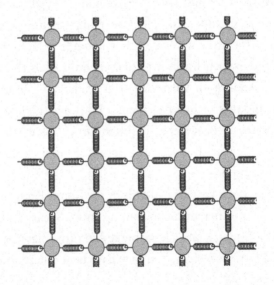

How a material responds to stress (the application of pressure) depends on how well the springs retain their springiness and how strong they are. Stretch a piece of Saran Wrap a little bit and it will return to its original shape when you let go. Stretch it a little harder: Although some areas won't return to their original shape, others will. If you stretch the Saran Wrap even more, you'll get two pieces of Saran Wrap.

When you pulled gently on the Saran Wrap, the springs between the atoms stretched a little. When you let go, the springs went back to their original lengths, so the Saran Wrap looked the same way it did before you stretched it. This is known as elastic deformation, which is reversible when the stress is removed.

When you stretched a little harder, you permanently stretched some of the springs and pulled other springs away from the atoms to which they had been attached. Some of those displaced springs may hook up with other atoms, but either way, they don't return to their original positions when you stop stretching. This is called plastic deformation, and it isn't reversible. Plastically deforming a material permanently changes its shape. When you pulled *really* hard on the Saran Wrap, you literally broke the springs between the atoms.

We need a metal that deforms plastically, but just enough so that we can shape the metal into tubes or fenders without breaking it. It's much like getting pie dough to just the right consistency: If the dough is too stiff, it will break when you try to roll it out. If it's too soft, it'll stick to the rolling pin and won't hold its shape.

Pure metals are extremely ductile. One cubic inch of pure gold can be flattened into a continuous sheet that is seven millionths of an inch thick and can cover a thirty-by-thirty-foot room. The problem is that very ductile materials are *too* easy to deform. A wedding ring made from pure gold would be so soft that just wearing it every day would destroy it. Pure iron has the same problem: It's not strong enough to stand up to a little nudge from the car behind you, much less the force of hitting a wall.

You have to understand what makes a metal weak before you can make it stronger. Imagine holding the bottom of a cube of iron on a table with one hand and pushing on the top of the cube with the other. As you push, the springs between atoms stretch. The more you push, the more the springs stretch, until you push hard enough that you break bonds, the atoms move, and new bonds form. This process, called "slip," happens many times, in different directions, and involves many planes of atoms until a flat piece of metal becomes a curved fender.

Slip usually begins at defects—mistakes in the ordering of the atoms. Defects are imperfections like missing atoms, atoms where they shouldn't be, or even an incomplete plane of atoms. Defects are ideal places for motion to begin, just like a hole in a sweater will start a run if you don't do something to prevent it.

The only way to stop slip is to prevent defects from moving. One way of keeping defects in place is to add even more defects, which is sort of like putting nail polish on a run in a pair of nylons. When the run reaches the nail polish, it can't run any farther.

Nature gives us another way to decrease slip by making it hard to grow perfect crystals. When you grow rock candy by evaporating a sugar solution, you get a bunch of small crystals of all different sizes that grow into one another. Most metals have the same type of structure, but the little crystals are so small that you can't see them with your naked eye. A high-powered microscope shows that, instead of one very-well-ordered crystal, a chunk of iron is usually made up of lots of smaller crystalline regions called grains. The size of the grains can vary, but in most metals they are on the order of the diameter of a human hair or less. A defect may start moving in one grain, but it stops when it comes to the end of the grain because it can't jump to the next one. It's like a run in a sweater coming to a seam. The boundaries separating the grains are another type of defect, but one that prevents the motion of other defects.

Creating defects in a metal is not difficult. Unbend one section of a paper clip, and then bend it back. Repeat this process a couple of times. The more times you bend the paper clip, the harder it becomes to bend. Each time you bend the paper clip, you create more defects. Those defects make the metal stronger.

Unfortunately (as you may have already realized), introducing defects isn't necessarily a solution to making a material strong—if you kept bending the paper clip as you read this, you probably now have a broken paper clip. Although defects can add strength, they also make the metal brittle—the opposite of ductile—and brittle metal can't be plastically deformed. Instead of stretching to adopt a new shape, a brittle material just breaks.

There are only ninety-two naturally occurring elements. None has the required combination of strength and ductility necessary for chassis construction; however, we can mix and match atoms from those ninety-two elements in different combinations to make literally thousands of different types of materials, each with different properties. Alloying is the process of combining two or more elements by intermixing their atoms until you don't have any regions that are all one type of atom.

Alloying gold with silver, nickel, copper, and/or zinc can make it stronger. Twenty-four-karat gold is pure gold, while 14-karat gold is fourteen parts of gold and ten parts of other metals. Twenty-four-karat gold is more expensive than 14-karat gold because there are more gold atoms in a gram of 24-karat gold than there are in a gram of 14-karat gold. Although 24-karat gold may be more precious, a 14-karat gold ring is much more likely to survive everyday wear.

The Roush Fenway fabricators use tubing made of steel, which is a general term for iron-based alloys. Soft, weak iron alloyed with other types of atoms—when done properly—creates a material that has just the right compromise between strength and ductility, making it ideal for chassis construction.

So how did people figure out which elements to mix with iron to make steel? Remember that early craftsmen used charcoal to extract iron from iron oxide. It turns out that the charcoal does more than just remove oxygen. Some of the carbon atoms from the charcoal get into the iron. Eventually, someone recognized that there was a correlation between how much charcoal was used (or how long the iron was left in contact with the charcoal) and how strong the iron was.

Wrought iron, probably the first iron alloy used, got its name because it is "wrought" (meaning worked) from the porous iron produced by heating iron oxide with charcoal. Wrought iron is much stronger than pure iron, which is surprising considering that wrought iron is less than 0.15 percent carbon by weight. Wrought iron that is 0.15 percent carbon by weight means that if you have one pound of wrought iron, only twenty-four thousandths of an ounce is carbon. An iron atom is a little more than four and a half times heavier than a carbon atom, so if you want to think of it in terms of numbers of atoms, there is one carbon atom for every 143 iron atoms.

If some carbon makes iron stronger, more carbon should make iron really strong. Cast iron, which has between 2 percent and 4 percent carbon by weight, is indeed stronger than wrought iron. The problem, as you know if you've ever cracked an engine block or broken a cast-iron frying pan, is that cast iron is very brittle. There's a tradeoff: Carbon makes iron stronger, but it also makes it more brittle.

So if no carbon means ductile but weak, and a couple percent carbon means strong but brittle, there should be some amount of carbon that makes steel that is strong but still ductile enough to be shaped. Mild or low-carbon steel—the specific type of steel used in chassis construction—has between 0.05 percent and 0.26 percent carbon by weight. One atom out of eighty-three in a piece of mild steel with 0.26 percent carbon by weight is a carbon atom. One of the fabricators explained to me that they use a specific steel alloy called 1018 in the chassis. The alloy number tells you what elements are present and what ratios they are present in. The alloy numbers for mild steels all start with "10." 1018 mild steel has between 0.15 and 0.20 percent carbon by weight and a bit of manganese thrown in to make the steel harder.

It is rather surprising that such a small number of carbon atoms changes soft iron into a material a driver trusts with his life. The improved properties come about because carbon changes the way the iron atoms arrange themselves. When you melt iron and carbon together, the atoms mingle freely; however, iron doesn't like carbon getting in its way when it tries to solidify. As the mixed iron-carbon liquid cools, the iron atoms push the carbon atoms out of the way and form grains of ferrite, a soft, ductile form of pure iron. The remaining liquid has a much higher ratio of carbon atoms to iron atoms. There eventually are some regions in the liquid where there is one carbon atom for every three iron atoms, and these regions solidify into thin sheets of cementite (Fe_3C), which is strong, but also very brittle. The end result is a Dagwood sandwich–like structure: alternating thin sheets of cementite and ferrite, one after the other, in each grain. The combination of layered cementite and ferrite is called pearlite, because it looks like mother-of-pearl under a microscope. Mild steel is a mix of pearlite and ferrite grains, and that mix is what makes this type of steel ductile enough to be formed into tubing but strong enough to protect the driver.

Many other types of steels are used in race cars, each having different combinations of atoms and therefore different properties. Suspension components need to be extra strong, so chrome-moly steels containing chromium and molybdenum as well as carbon are used for these parts. These alloys have three to four times the strength of mild steel. The tradeoff is that they are much more difficult to shape and weld, so these alloys are used mostly for parts that can be bolted onto the car.

You can tailor the properties of steel by changing the numbers and types of atoms, but it also is possible to get different properties from two steels with the exact same numbers and types of atoms if the atoms are arranged differently in the two materials. Cooling a liquid mixture slowly allows the atoms time to get into their preferred positions. If you cool the same mixture quickly, the atoms can be frozen in positions other than the ones they would occupy if they had been cooled slowly.

Regardless of how they are made, all mild steels are magnetic. NASCAR takes advantage of this characteristic during inspection. A magnet will stick to mild steel, but it won't stick to "exotic" metals like aluminum or titanium alloys. These alloys are strong enough for chassis construction and lighter than steel, but they can be much more expensive. To prevent costs from skyrocketing, NASCAR does not allow the use of these alloys.

As we continued through the shop, Jamie told me that each Roush driver gets the same equipment. The partially completed chassis sitting in front of me could become a No. 26 car or a No. 99 car; however, that doesn't mean that all the cars built in the Roush shop are identical. In 2001, NASCAR started a major research initiative to develop a new race car that would keep the driver safer, make racing more competitive, and decrease the cost of ownership. This initiative produced a vehicle originally dubbed the "Car of Tomorrow." This car ran short tracks, road courses, and the second Talladega race in

2007 and then became the only NASCAR race car in 2008. I visited the shop at a time when teams were producing both cars.

I could tell that the stack of frames piled on the floor in front of me were for the old version of the car, because they were labeled according to the type of track on which they were to run. NAS-CAR tracks can be divided into four categories: road courses (Sonoma and Watkins Glen), where the car must turn right as well as left; superspeedways (Daytona and Talladega), which are the longest and most steeply banked tracks; short tracks (like Martinsville, Richmond, and Bristol), which are a mile and under; and intermediate tracks such as Texas, Las Vegas, and Atlanta. Different chassis designs are built to optimize a car for the different types of tracks. For example, the high speeds and steep banking of Daytona and Talladega place tremendous stresses on the car, so the chassis is stiffened. A chassis designed for a short track like Bristol, where bumpers are part of your race strategy, might have extra bracing in the front and back. Most teams have taken this customization even further, making different chassis for different tracks in the same category.

This pile of frames is symbolic of one of NASCAR's motivations for introducing the new car. In contrast to the old chassis, Roush Fenway makes exactly one type of Car of Tomorrow chassis, and it's the same design every other team makes. NASCAR specifes the position, size, wall thickness, and material of every single piece of tubing, which means that teams don't have to—or don't get to, depending on who you're talking to—specialize the chassis to different tracks. The same chassis that races at Daytona will race at Martinsville and Watkins Glen. The advantage, according to NASCAR, is that an owner should only need to build eight to ten cars per team instead of the fifteen to twenty they might build now.

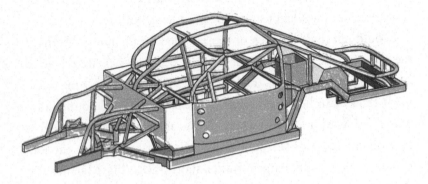

Jamie explained that building a complete chassis like the one shown above takes anywhere from 170 to 250 person-hours over the course of two to three weeks. The chassis for both styles of cars are built in three main parts: a central section consisting of a frame-rail base and a roll cage that surrounds the driver, and the front and rear "clips." The front clip, which supports the engine and the front suspension, is designed to push the engine downward instead of into the driver's compartment in a frontal crash. The rear clip supports the rear suspension and has a tube frame to protect the fuel cell.

The foundation of each section is the frame rail: rectangular tubing upon which the rest of the structure is built. The strongest tubing—the tubing with the largest diameter and the thickest walls—is used closest to the driver. Tubing size decreases the farther away from the driver you move. For example, the frame rails of the center section are rectangular, three-by-four-inch mild steel tubing with 0.125" wall thickness. The frame rails for the front and rear clips are made from two-by-three-inch rectangular mild steel tubing with a slightly thinner wall diameter of 0.083".

The center section's most important function is to protect the driver. The left-side frame rails, which are closest to the driver, are doubled for extra strength. The piece of tubing that runs vertically from the roof down the center of the windshield, called the "Earnhardt bar," reinforces the roof and prevents large objects from entering the driver's compartment.

Many changes on the new car are the result of accumulated experience. The driveshaft is a large, heavy piece of metal that runs under the car from the transmission to the rear wheels and rotates at speeds of up to 10,000 revolutions per minute. A loose driveshaft could drag on the ground and spark, or even come off entirely and pose a hazard for other cars. Precautions had been taken in the old car: In addition to painting the driveshaft white to make it visible on the track, a steel hoop encircled the driveshaft to keep it from dragging if it came loose. The new car takes safety one step further by mandating a 360-degree tunnel structure that completely encloses the driveshaft.

Some of the most significant changes come directly from NASCAR's safety research. The old door had four horizontal door bars aligned in a single plane, while the new car features graduated door bars: The top bar sticks out farther than the next lower bar, which sticks out farther than the bar below it and so on. The staggered configuration allows the bars to crush sequentially instead of all at once.

The engine exhaust tubes exit from the right side of the car, so that when the cars pull onto pit road, the exhaust is pointed away from the pit wall. The exhaust path in the old car ran underneath the driver's seat, which increased the cockpit temperature—and the likelihood of heat stress. The new car routes the exhaust under the right-side frame rails.

In the shop, chassis subsections that are under construction are often perched on wheels that look like those on a grocery-store shopping cart, so that workers can move the large metal structures easily. As we watched a fabricator wheel a front clip over to a roll-cage assembly, Jamie explained that the roll cage and clips are constructed separately and then welded together.

Welding is one of the most important jobs in building the chassis: It doesn't matter how strong the steel tubing is if the pieces come apart in a crash. A good weld should literally turn two metal pieces into one by allowing atoms from one piece of metal to mix with atoms from the other piece of metal. That mixing won't happen just by pressing two pieces of solid metal together.

The atoms in a solid stay close to their assigned spots, just like the fans in their seats during a race. Like race fans, the atoms aren't sitting totally still. Atoms in solids vibrate about their assigned spots in the crystal, much like race fans move around even though they are sitting in their seats. The atoms, however, oscillate a lot faster, moving back and forth literally a hundred million million times per second.

The temperature of a solid reflects how much the atoms are vibrating. As the material gets hotter, the atoms vibrate more vigorously and push each other a little farther apart than when they were cold, which makes the solid a little larger. This expansion isn't big enough to see with the naked eye, but it is enough that running a glass jar under hot water lets you remove its stubborn metal lid. Metal expands more than glass when both are heated by the same amount. The metal lid expands slightly more than the glass jar, which lets you remove the lid.

As the temperature continues to increase, the atoms vibrate more and more. Eventually, some atoms on the surface break the bonds connecting them to neighboring atoms and move away from their assigned positions. This change from solid to liquid is similar to what happens when the race ends: People leave the ordered pattern of their seats and a fluid flow of fans heads toward the parking lots. Once there is some liquid at both surfaces, the atoms from the two pieces can intermix. When they cool, they should look like a single piece of metal. Before joining the two pieces, the welder meticulously cleans both metal parts. Dirt and grease between two pieces prevent the atoms from mixing together, and the dirt or grease molecules weaken the weld joint.

Early welding techniques weren't very effective because most metals have to get very hot to melt. A smith would get the metals as hot as possible and then literally had to beat the two pieces together to make the atoms intermix. The Roush Fenway welders use electrical arcs to heat the steel tubing in a controlled manner. The welders also add extra metal (called "filler") to the area where the pieces meet to make stronger joints. When joining two components of the frame, for example, the welder heats the two pieces and the filler metal, as shown below (left). The surfaces of the metals become liquid and mix (middle), and then finally solidify (right). If the weld is good,

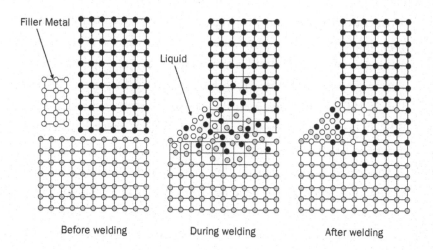

Filler Metal

Liquid

Before welding   During welding   After welding

you can't tell where the first piece ends and the second piece begins, making the chassis literally one continuous piece of steel.

Although the chassis is made mostly of steel tubing, some sheet metal is added during its construction. Firewalls are thin metal panels about thirty-one thousandths of an inch thick that separate the engine and the fuel cell from the driver's compartment. Firewalls keep flames and heat away from the driver in the event of a fire. The floor pan, which forms the bottom of the driver's compartment, is welded into the base of the center frame. A metal box is constructed in the rear clip to protect the fuel cell. Brackets are attached for parts that will be mounted later, such as the engine, seat, and fuel cell. Some of the transmission and suspension, as well as components like the pedals and dashboard, may be installed now, Jamie explained, because fabricators have better access now than they will after the body is added.

Jamie's point was illustrated as we headed back toward the lobby, passing through an area where team members were putting the final touches on the insides of some almost-complete cars. A pair of legs sticking out a window, or a person upside down in the driver's compartment trying to get just the right angle to attach a bolt, are common sights in the shop. Jamie mentioned that the car with the legs sticking out was one of Roush's first complete new cars and was scheduled to go over to NASCAR to be checked against specs the next day. NASCAR certifies each chassis that passes inspection with radio-frequency identification tags that will be checked when the car arrives at a track to race.

The legs disappeared from the window and the young man attached to them poked his head out from the cockpit of a No. 99 car that would eventually be driven by Carl Edwards. He asked if one of us would pass him some rivets sitting on a cart just out of his reach.

"It takes a really long time to get in and out of these cars," he explained as I handed him the container of rivets. Jamie smiled.

"If anything goes wrong with this car, you know, we're blaming you," he said, laughing. I asked him for the chassis number—which identifies the car—before departing. If the car wins, I'm expecting a visit from Carl for a personal thank-you.

Two

Skin Care

If you've never been to a race shop, let me warn you that most are pretty underwhelming—at least on the outside. It was something of a letdown to find that most shops are situated in unassuming business parks. Hendrick Motorsports—the home of Jeff Gordon, Jimmie Johnson, Casey Mears, and Dale Earnhardt, Jr.—is located just across the expressway from the Roush Fenway shop. HMS is one of the larger NASCAR enterprises, with four NASCAR Sprint Cup teams, two NASCAR Nationwide Series teams, and about 550 employees. The twelve buildings (four of which are open to the public) form their own ninety-acre business park.

The Hendrick Museum documents the history of Hendrick Motorsports over the last two decades and gives you a close-up view of some historic cars. The morning I visited, Jimmie Johnson's winning 2006 Daytona 500 car had just been put on display, complete with confetti still stuck to it.

One reason I like the Hendrick Museum so much is that their displays give you a good view of the inside of the car: Engine blocks, carburetors, oil tanks, and brake rotors carry labels and explanations to show you which part goes where. A complete chassis, similar to the one we left at Roush Fenway Racing, sits in one corner of the museum, and a body in various stages of completion—sheet metal, primer, and paint—is suspended over it.

The process of putting the sheet metal on the car is called "hanging" the body, and HMS gave me an inside look at the process. I walked across the parking lot to the administration building to meet Megan Whiteside, media relations specialist for HMS. The Hendrick campus is large enough that we had to take a golf cart over to the No. 24 DuPont and No. 48 Lowe's race-shop building.

If the outsides of many race shops aren't impressive, the insides certainly make up for it. Two-story-high glass walls form the front of the building. Wood paneling in the lobby sets off glass cases packed with trophies won by Jeff Gordon and Jimmie Johnson. Megan pointed out the public viewing area: It's designed like a track, with a catchfence separating visitors from a car-storage area and, just beyond that, the main shop.

Returning to the lobby, we met Mark Thoreson, who would give me a tour of the No. 24/No. 48 shop. We were accompanied by Larry Deas, the manager of DuPont Motorsports, which has sponsored Jeff Gordon since he started competing at the NASCAR Sprint Cup Series level in 1993. Mark has worked with the No. 24 team since 1998, serving as team manager for the 2005 and 2006 seasons. He is the link between the crew chiefs and the car builders. He oversees ordering equipment and managing people, and he has the ultimate responsibility for ensuring that cars coming out of the shop meet the standards of Hendrick Motorsports and NASCAR.

The Hendrick fab-shop walls are light gray, but the large rooms are brightly lit and the high ceilings make them feel even larger. Stations in each room hold cars in various stages of completion. Fabricators—the people who specialize in shaping sheet metal and attaching it to the chassis—buzz back and forth between cars and tools, practicing an art that goes back as far as humans have known how to extract metals from the Earth.

You can find examples of sheet metalworking from thousands of years ago in natural-history museums. Most of these artifacts were made by hammering a flat piece of metal to the desired shape. While this made interesting textures for decorative items, the irregular surfaces were by no means aerodynamic, and race cars are all about being aerodynamic.

Most metal pieces on production cars are made by placing a flat metal sheet into a hydraulic stamping machine that, by sheer force, molds the metal to the shape of the die in which it sits. A stamping machine can churn out piece after piece, stopping only to be restocked with raw metal. If you expected a production line churning out fenders at Hendrick Motorsports, the relative quiet of the shop would disappoint you. This facility is designed for craftsmanship, not for volume.

With the exception of the hood, roof, deck lid (aka the trunk), and bumpers, every piece of the race-car body starts as a flat piece of sheet metal. Some pieces are stamped into rough shapes and then finished, but the majority of the body is made entirely by hand. Fabricators hold flat pieces of metal against the chassis and mark rough cut lines with a Sharpie. Metal snips and shears make a first approximation of the desired shape, and then the piece is taken to the English wheel.

The English wheel was developed in the early 1900s to make smooth curves in sheet metal. An English wheel essentially is two metal rolling pins a few inches in width set one on top of the other. The upper wheel, which can vary from about three to nine inches in diameter, has a flat surface, while side-to-side curvature in the lower wheel allows shapes of different radiuses to be formed.

An experienced fabricator makes using the English wheel look simple. He places a piece of sheet metal between the two wheels and adjusts the pressure on the top wheel. When he rolls the metal back and forth between the wheels, the pressure from the top wheel makes the metal thinner in places, causing it to curve. The fabricator runs the metal back and forth quickly between the wheels, periodically checking the piece against the car for fit. A talented fabricator can turn a flat piece of metal into a fender in just a few hours.

It takes a special type of steel to stand up to the rolling and shaping without breaking. Hendrick uses a mild steel called CR1008DQAK. "CR" stands for cold-rolled, which means that the sheet stock is made by passing a piece of steel between two very large rollers until it reaches the desired thickness. "1008" identifies the type of steel, which is mostly iron with about 0.10 percent carbon and 0.3 to 0.5 percent manganese by weight. There is less carbon in this steel than in chassis steel. The body panels don't have to be as strong as the chassis, but they do have to be more ductile.

"DQAK" is short for "Drawing Quality Aluminum-Killed." Drawing is the process of shaping a material by forcing it through a hole or rolling it. Bending or rolling metals creates defects, as we saw in the paper-clip experiment in Chapter 1. "Drawing Quality" means that this steel is made specifically to withstand extreme pressing, drawing, or other forming processes without creating a lot of defects.

The "Aluminum Killed" part describes the processing method. Steel is made by melting elements together according to a particular recipe and then cooling the mixture until solid. Oxygen and nitrogen atoms from the air can become dissolved in the molten steel. When the steel cools, these atoms form defects that make the steel brittle, or they react with carbon in the molten steel to form carbon dioxide gas bubbles that can produce holes in the cooled steel.

Oxygen and nitrogen atoms would rather bond with aluminum atoms than carbon atoms. Aluminum in molten steel forms compounds with oxygen and nitrogen atoms, preventing them from becoming defects in the solid steel. For example, aluminum and ni-trogen form small grains of aluminum nitride, which actually helps the steel grains organize themselves for maximum ductility. The term "killed" arises because metals with dissolved gas bubbles make noise when poured into a mold, just like carbonated water makes noise when you pour it into a glass. Molten aluminum-killed steel has no trapped gas bubbles, so it pours quietly.

Mark mentioned that they do as much of the work on the chassis as possible prior to hanging the body. When I asked why, we knelt near the front wheel opening, and Mark suggested that I—gently—feel the front fender. I was surprised at how thin the sheet metal is: NASCAR requires a minimum 24-gauge thickness, which is about twenty-five thousandths of an inch. The metal thickness varies a little bit along the car because making a curve in the metal means making the metal thinner in that spot. Even at its thickest, the skin is surprisingly flexible. Mark showed me how braces be-tween the chassis and skin provide additional support for the body.

With the metalworking tools and expertise in the shop, almost any body shape can be made; however, NASCAR imposes two pri-mary constraints. The first is that the roof, hood, and deck lid must be provided by the manufacturer. The second set of constraints hangs on the wall: the templates.

Templates are metal pieces that fit over specific contours of the car. They are used in the shop during construction, to check the car before taking it to the track, and by NASCAR at the track. The old car was checked with a set of thirty-four independent tem-plates. The longest of these, called the overall template, fit along the midpoint of the car from nose to tail. The spoilers were installed in two pieces, with a small gap in the middle that was taped up after the overall template had been checked. Thirty-three other templates checked dimensions such as the width of the hood and roof, and the slope of the windshield. Some templates applied to all cars, while others were specific to the manufacturer.

The rows of cars in the public viewing area—all were the "old" version—are a testament to how much you can pull, push, warp, and otherwise mangle the old car and still have it fit its templates. Look at a picture of a non–road course old car taken head on. It is shaped like a kidney bean, with the front and the back curving toward the car's right side. Compare how much of the left-front fender you can see relative to the right-front fender. This offset had become so extreme that Gary Nelson, past vice president of the NASCAR Research and Development Center, said that the car looked like it had been in an accident before it even got out on the track.

The twisted shape of the old car is the result of intensive aerodynamics research. Teams learned that small changes in body shape can produce advantages that make the difference between winning and losing. Teams started customizing the old car's body not just to the *type* of track, but to each individual track. Mark counted off a typical inventory for one team: two road-course cars, three superspeedway cars, three or four short-track cars, and five to six intermediate-track cars. Hendrick might build six or seven new cars per team each year—more for a driver who doesn't always return cars in the same shape as when they left the shop.

The new car was developed to decrease the total number of cars each team needs because there is exactly one allowed body shape. Mark thinks that Hendrick likely will decrease their inventory from thirteen or fourteen cars per team to ten cars per team in 2008, although he adds skeptically, "We'll see." When I asked Mark how much they can tweak the new car's body, he replied somberly, "Not at all" and pointed to a large piece of paper hanging on the wall across the shop.

When NASCAR referred to a "blueprint" for the new car, I

thought it was a metaphor, but the piece of paper to which Mark pointed literally *is* a blueprint. In addition to changing the car, NASCAR also made important changes in the inspection process. The thirty-four independent templates used for the old car defined mostly one-dimensional contours. A few templates had foldout arms, but most were long, linear pieces of metal fit sequentially to the car.

Instead of a series of independent templates, the new car must fit three much larger sets of interlinked templates, called "grids." The largest grid fits over the top of the car from hood to tail and side to side. This grid hangs from the top of a sturdy steel scaffold, making it look like a giant metallic spider poised to drop down and engulf the car. The main grid is the same for all cars, with the manufacturer's identity coming from the nose and rear-end grids, both of which also incorporate multiple templates into a single unit.

It doesn't seem like it should make a difference whether templates are applied one after another or all at the same time. Let's say you need to drive from your house to the bank across town. You can choose from a number of different routes, all of which get you from point A (your house) to point B (the bank). Then your spouse asks you to stop and pick up a gallon of milk at the convenience store. That request limits your choices: You now have to choose a route that takes you past the bank *and* a convenience store. Then you remember that you need to pick up some dry cleaning, which eliminates any routes that don't go past the bank, a convenience store, and the dry cleaners. Each added constraint decreases the flexibility you have in picking your route. The new template grids are, to the teams, like having someone not only tell you where to stop, but specifying the exact route you should take.

The fixed body of the new car eliminates any aerodynamic adjustability in the body. The spoiler on the rear of the car is replaced by a wing, and a front splitter—a shelf on the very bottom of the front

nose piece—is added. These two changes shift all of the aerodynamic adjustability from the body, which can't be changed at the
track, to the splitter and wing, which can.

Mark pointed out some other differences between the old and new
cars, which you can see in the drawing below. The new car's body is
set farther forward on the chassis relative to the old car. The new car
is more snub-nosed but has a longer rear end. The body is now cen-
tered on the chassis left to right, although the chassis has a built-in
one-inch offset to the left. The rear quarter panel is offset two inches
to the right. The windshield is more upright, and both bumpers are
much more square, which led Tony Stewart to dub it "The Flying
Brick."

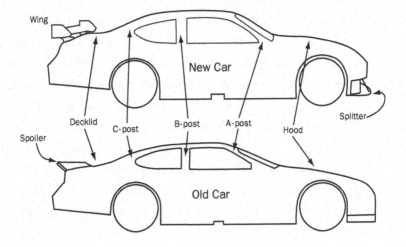

The new car gives the driver a little more space. It is two and a
half inches taller. The roof increased from 51 inches to 53½ inches,
which should make racing much more comfortable for drivers like
six-foot-two Dale Jarrett. The new greenhouse—everything above

the window ledge—is four inches wider. The driver's seat has been moved closer to the centerline of the chassis, which moves the driver's head farther away from the left side of the roll cage. The window openings are two inches taller than they used to be, enabling drivers to get out of the car faster in an emergency. The wheelbase—the distance between the center of the front wheel and the center of the back wheel—remains 110 inches.

Mark explained that, despite the differences in configuration, bodies for both types of cars are constructed similarly. After being shaped, each piece of sheet metal is welded to adjoining pieces of the body. The roof is usually the first part to be placed, followed by the deck lid and hood. The greenhouse, which consists of the A-, B-, and C-posts, door tops, and the roof, is added next. When teams turned production cars into race cars, the doors had to be strapped or welded shut to prevent them from popping open in a wreck. Since today's cars are built from scratch, there's no point putting in doors, so sheet metal covers the entire side of the car. Fenders and quarter panels follow, with every addition checked against the templates. Crush panels—sheet metal extending from the firewall to the fenders—form a metal seal around the cockpit and keep heat, fumes, and debris out of the car.

When I asked why the front and rear bumper covers are a darker gray than the rest of the body, Larry Deas, the manager of DuPont Motorsports, explained that these are the only two parts of the body not made of steel. The front and rear bumper covers are a molded Kevlar-reinforced composite. Kevlar is a DuPont polymer with five times the strength of an equal weight of steel.

Poly means "many" and *mer* means "unit." A polymer is a series of units linked together, just as a train is a series of boxcars linked together. Polymers are found in nature (cellulose, spider silk, and natural rubber, for example) and made in the lab.

Polyethylene is a fairly simple polymer that is used in everything from sandwich bags to water pipes. Polyethylene literally means "many ethylenes." Polyethylene is a chain of CH_2 groups linked together. I've drawn only a few ethylene groups, but real polymers can have thousands or millions of units.

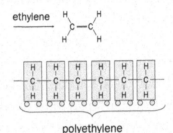

polyethylene

The polymer "backbone" is the chain of atoms forming the main structure. Sidegroups, which are atoms or groups of atoms, extend from the backbone. The backbone of polyethylene is the chain of carbon atoms and the sidegroups are hydrogen atoms. Other polymers have more complex side groups and/or backbones.

Stephanie Kwolek, a DuPont chemist, is credited with the major discovery that led to Kevlar. Concerns about a possible gas shortage in the mid-1960s spurred scientists to study ways to make cars lighter without compromising safety. At the time, tires were reinforced with heavy steel wire. DuPont wanted to develop lighter fibers with enough strength to replace steel.

Kwolek's job was to mix chemicals to make a polymer and then dissolve the polymer in a liquid so that a coworker could spin the liquid into a fiber. In 1964, a polymer called para aminobenzoic acid surprised her: The solution was opaque, unlike every other polymer solution she had studied. The fiber spinner initially refused to spin her solution because he thought it would clog up his machine, but he eventually relented. When the fiber's strength was tested, Kwolek was so surprised at the results that she asked for the tests to be repeated. The additional tests confirmed the results: These fibers were unexpectedly strong, even at high temperatures.

The problem was that this polymer was too expensive to develop into a commercial product, so DuPont scientists launched a research effort to understand *why* Kwolek's polymer had such exceptional properties. Once they understood what made the polymer different, they could develop a less-expensive polymer with comparable properties.

Kevlar is a member of the aramid family, which is a group of polymers with straight, rigid molecules. Its strength is due to the long polymer chains that align with each other and bond tightly together. The fibers can be used by themselves, woven into cloth, or, as in the case of the nose and tail pieces, be part of a composite. Composites are two or more materials that form a single unit without either component losing its identity. Most composites have some type of reinforcement (fibers, particles, or fabric, for example) surrounded by a matrix. Steel rebar–reinforced concrete is a composite. The concrete is the matrix and the steel rebar is the reinforcement.

The front and rear fascias of the race cars are made by placing layers of Kevlar fabric in a mold and saturating them with a resin—a thick, sticky liquid that dries to a solid. The resin penetrates the fabric and hardens. Kevlar-reinforced fascias are significantly lighter than metal parts, but just as strong.

The final step is fine-tuning the shape with body filler, which you might know as Bondo. Body filler is to a race car as spackle is to home repair. Like spackle, body filler goes on as a paste and dries to a solid. The ability to apply it as a paste allows the fabricator to build up areas on the car to match the templates.

Body fillers are plastics. Plastics come in two general types. Thermoplastics can be formed and re-formed. A thermoplastic polymer like polyethylene can be heated and shaped in a mold. If you want to change its shape, you can melt it and remold it. Body fillers are thermoset polymers, which, once formed, are difficult to reshape. The atoms in different polymer chains form chemical bonds with each other, creating a three-dimensional network that is much stronger than the unconnected polymer chains. This bonding, which is called "crosslinking," isn't easy to undo once it is done.

Crosslinking can produce big changes in a material. Natural rubber is soft, but crosslinking makes it strong enough to use in tires. Polymer crosslinking plays an important role in perming— chemically curling or straightening hair. Hair is a polymer made up of protein chains called keratins that are crosslinked with sulfur-containing groups. (The sulfur atoms are responsible for the nasty smell of burning hair.) Perm solution breaks some of the crosslinks in your hair. The neutralizer re-forms the crosslinks, and if your hair is in curlers, the new crosslinks lock in the curl.

Because the crosslinking reaction is irreversible, thermoset polymers like body filler come in two separate parts: a filler and a catalyst, or hardener. The filler contains a polymer and a reinforcing material, such as small fibers. When the catalyst or hardener is first added to the filler, the compound is flexible enough to be shaped. The fabricator mixes the two parts of the body filler and applies it to the car. As the mixture cures, crosslinking starts and the compound hardens. If an area is too high, the fabricator sands the filler until the template fits. Layers of body filler are added to build up areas that are too low.

Even with the flexibility of Bondo, NASCAR gives the team some tolerances in satisfying the templates. The old templates allowed the car to be from 1/16" to 1/2" off the mandated dimensions. The new car must satisfy a tolerance of 1/4" at most places. Mark pointed out that tolerances are necessary not just from a manufacturing standpoint, but because the car's dimensions can change as it gets warmer sitting in the sun.

Tighter tolerances require more precise measurements. Let's say you need to check the height of your rear quarter panel. You can measure the distance from the garage floor to the bottom of the panel, but if the floor is uneven, you'll get a different value depending on exactly where the car is located when you measure.

NASCAR teams use a surface plate—a very flat steel plate large enough to hold the entire car—to improve the precision of their car building and suspension setups. Mark explained that the surface plates at Hendrick have less than seven thousandths of an inch difference in height between any two points. A really good-quality trash bag is about 3 mils thick (a mil is a thousandth of an inch), so that's a difference in height of about the thickness of two trash bags over a twelve-by-twenty-foot area. Most surface plates are made of a single piece of precision-milled steel or sometimes granite; however, surface plates can also be made of epoxy.

Epoxy is a thermoset polymer that starts as a liquid and hardens as it crosslinks. You may have heard that liquids always find their own level. The atoms in a liquid are much less connected to each other than the atoms in a solid, so the surface of a liquid remains flat. If you put water in a glass, the water surface remains level regardless of how you tilt the glass. Liquid epoxy for a surface plate is poured into a form and self-levels. Epoxy surface plates claim as little as 3 mils variation across the entire plate when professionally installed.

There is a limit to how precisely you can measure something with a ruler or tape measure. The smallest division on a steel ruler is 1/32" or maybe 1/16" (the latter of which is about 62½ mils). As the divisions get smaller, your eye can't see the spaces between them. Increased precision requires measuring with something other than your eyes.

A digitizing arm is a long, articulated arm with a pointer at the end that looks like a finger. You may hear this device called a Faro arm or a Romer arm, after two popular manufacturers' models. After selecting a reference point, you touch the pointer to the car and the digitizing arm outputs three coordinates that tell you how far across, over, and up the pointer is from your reference point, within a tolerance of up to 1 mil. By touching the pointer to different locations on the car, you can get enough coordinate sets to feed into a computer and build a very accurate virtual model of the car.

Laser scanning is another technology that is increasingly being used. Instead of physically touching the car, a laser line, usually a few inches long, is projected onto the surface. A detector at a known distance captures the reflected laser beam, and a computer uses trigonometry to calculate the coordinates of the line. Some teams laser scan every car before it hits the track.

It takes around 240 person-hours—three people over about ten days—to take the car from a steel skeleton to a full-bodied race car, although Mark noted that the shop can produce a complete car from scratch in as little as a week. "Not that we'd be real happy at the end of that week," he added.

A car leaving the body shop doesn't do justice to the many hours of shaping, pushing, pulling, forming, welding, and Bondo-ing that went into it. It looks like an automotive Frankenstein with its gray bumpers and splotches of body filler over the dull sheet metal. The good news is that a full makeover is just around the corner.

Three

Makeup

Walking out of the Hendrick Motorsports fab shop, I caught a glimpse of a car coming out of a paint booth. Even without decals or numbers, I knew it was a Jimmie Johnson car. We recognize the cars we cheer for first by their color. Before your brain registers a logo or a number, you recognize Lowe's Blue or Budweiser Red. The same sunlight hits the No. 48 and the No. 8, so why does one look blue and the other red?

"White" light actually contains many different colors, which you can see by shining white light into a prism. When the light emerges on the other side, it's been separated into a rainbow of color. You may have learned the mnemonic ROY G BIV (red orange yellow green blue indigo violet) for the order of the colors in a rainbow, which makes it sound like a rainbow is made of seven distinct colors; however, a rainbow is really a continuous band with many different shades of each color. Light bends when it enters a prism because it slows down. It's harder for light to travel through glass than through air. Different colors of light travel through the glass at different speeds, so some colors of light bend more than others. Violet light travels more slowly than red light, so violet light bends more than red light.

Some objects, like the sun, emit their own light, but most—including race cars—reflect light coming from another source.

Sunlight hits a car on the track and you see the sunlight reflected from the car. The sunlight hitting the No. 48 contains all of the colors of the rainbow, but the car only reflects some of those colors back to your eyes. The paint absorbs all of the colors that make up white light *except* those that comprise the characteristic blue of the No. 48 car. Those colors are the ones that reflect back to your eyes. Which colors are absorbed and which are reflected is determined by the pigments in the paint. Different pigments absorb and reflect different colors of light to different degrees, depending on what types of atoms are contained in the pigment.

The first pigments came from minerals, plants, and animals. Purple pigment, for example, was made as early as 1200 B.C. by the Phoenicians from snail mucus. Collecting enough snails to dye an entire garment was so time-consuming that only the wealthy could afford purple clothes. We learned how to make artificial pigments in the early 1700s, which made it possible for anyone to wear any color—or to put it on a car, once cars were invented.

We understand pigments so well today that you can take a towel to Lowe's and bring home paint that matches your towel exactly. The device that makes this possible is called a spectrophotometer. "Spectra" means color, "photo" means light, and a meter is a device that measures "how much." The spectrophotometer shines white light on your towel and measures how much of each light color is reflected. A computer translates that information into what types of pigments and how much of each are needed to make paint that reflects the same colors of light as your towel.

Once you know how to make the color pigment you want, you need a way to get—and keep—the pigment on the car. Even snail mucus is not sticky enough to stand up to 180 mph. Pigments have to be suspended in a binder that allows you to apply the pigment and then holds the pigment on the car.

Water-based paint, like latex house paint, has a polymer binder and pigment suspended in water. You brush or roll the paint on the wall, and when the water evaporates, the polymers hold the pigment to the surface like bungee cords hold down a load in a pickup truck. This, incidentally, is why it is difficult to remove dried paint: The drier the paint gets, the stronger the bonds become. Latex paint is easy to apply and (before it dries) clean up with water; however, latex paint doesn't adhere well to metal and doesn't stand up to scrubbing—much less brushing the turn-2 wall at Darlington.

Cars—including those waiting to get into the paint booths at Hendrick Motorsports—use acrylic urethane paints. Urethanes are molecules with a specific arrangement of nitrogen, hydrogen, carbon, and oxygen atoms. Unlike latex paint, acrylic urethane paint comes in two parts. The first part is an acrylic-based resin that contains the pigment and a polymer, and the second part is a hardener or catalyst. Neither part is paint by itself—they don't become paint until they are mixed and a chemical reaction occurs.

Because of the molecules involved, acrylic-urethane paints can't use water as their base liquid. Automotive paints usually have to use volatile organic compounds (VOCs)—chemicals like acetone, alcohols, and toluene, which evaporate easily. You can usually identify VOCs by their smell: Volatile means that it is easy for molecules to escape the liquid and make their way to your nose. VOCs are produced artificially (you can find them in some spray paints, paint thinners, and nail polish removers) and naturally (in trees and cow farts).

The two paint components are mixed together just before being applied to the car with a large airbrush. Paint is poured into a reservoir and air is drawn into the paint gun. The air is squeezed through a narrow tube, which makes the air move more quickly. The faster the air flows, the lower the pressure it has. Paint is sucked into the low-pressure region and the high-speed air jet breaks the paint into tiny droplets that are carried with the air to the car's surface. Applying the paint in small droplets allows a very even application and a smooth blending of colors.

Acrylic-urethane paint "dries" in two stages. First, the VOCs in which the pigments and binder were dispersed start to evaporate. The second step is crosslinking, where the acrylic-urethane polymers form chemical bonds with each other. The resulting polymer network has a hard, glossy finish.

Paint booths are the size of drive-in car washes. Turntables are used to rotate the car onto its sides or roof so that every surface can be painted. Heaters raise the booth temperature to between 120°F and 150°F, which helps the paint cure faster. The booths are airtight, so that dust and dirt from the shop can't get in and the paint and the VOCs don't get out. Air flows into the booth from the top, pushing excess paint toward the floor, where it falls through grates and into special filters.

Keeping the VOCs from getting into the atmosphere is important. For example, methane (a primary component of cow farts) is a greenhouse gas that contributes to global warming. Other VOCs create ozone or help methane stick around longer, both of which decrease air quality. Some VOCs are suspected carcinogens and some are just plain nasty-smelling—all good reasons why you should use spray paint only in well-ventilated areas. Hendrick Motorsports painters—like those at all NASCAR teams—wear respirators when prepping the cars hoods that supply fresh air when they paint the cars. Mark noted that Hendrick employees are tested randomly each year to ensure that their safety precautions are effective.

Larry explained that DuPont is developing paints that are friendlier to people and the environment. The challenge is developing paints with fewer VOCs that still provide the shiny finish expected on a new car—or race car. Since Los Angeles already faces significant air-quality challenges, it is appropriate that Jeff Gordon featured one of DuPont's new low-VOC paints on his car at both 2007 Fontana races. DuPont Cromax Pro paint not only sponsored the car, but was used to paint it as well. Cromax Pro is a waterborne acrylic urethane paint that has nearly 50 percent fewer VOCs than traditional acrylic urethane paints.

Mark took me through the steps involved in painting the car. Anything that shouldn't be painted and can be removed easily is detached from the car. The car is pressure-washed and then etched with an acid rinse that helps the paint adhere better. The car is baked at around 150°F for an hour to remove any water or other moisture left by the washing process. Anything that can't be removed but shouldn't be painted has to be covered. Time is at a premium, so Hendrick painters don't use tape: They have clear decals that are just the right size and shape to cover the areas that need to be kept paint-free.

NASCAR mandates that the interiors of the cars be painted a light color. "DuPont Dove Gray is what we use," Larry told me. The only exception to the gray interior is the inch-plus-diameter pipe housing the fuel line, which is painted red to help emergency crews identify it quickly.

The dashboard and the rear shelf of the car ("where you would put the speakers," Mark explained) are painted with a black "wrinkle paint" that goes on smooth and crinkles up as it dries. Rough surfaces don't reflect light as well as glossy finishes. The wrinkle paint reduces glare and makes it easier for the driver to read the gauges in the dash.

The first exterior coat is a gray primer that smoothes uneven surfaces and makes a slightly rough surface to which the top coat can adhere. Rougher surfaces give the paint more places to form bonds with the car body. You can think of primer as a sort of double-sided sticky tape that forms good bonds with the body on one side and with the final paint coat on the other. The car is then baked to cure the primer.

After the car cools, the templates are checked again to make sure there hasn't been any distortion during the heating and cooling cycles. If any more tweaking is necessary, body filler is applied and primed, and the car is baked again. One final check against the templates and it's time for the final paint coat.

You still can't tell whether this car will be Jimmie's or Jeff's because the preparation of every car is the same to this point. The paint scheme—the design that gives the car its identity and its style—starts with the color application. A paint scheme is a high-speed advertisement that has to look good while satisfying the sponsor and NASCAR rules. Joyce Julius & Associates, a marketing analysis company that reviews every race broadcast to tabulate how much time each sponsor's logo spends on the screen, estimates that a car finishing in the top 25 at the 2006 Daytona 500 got the equivalent of about $7 million worth of thirty-second commercials.

Each car has a "standard" paint scheme, but most cars run special paint schemes for special occasions or because the car's sponsorship is split between two or more companies. For example, Jeff Gordon ran a green DuPont/Nicorette paint scheme for the 2007 Las Vegas race instead of his standard blue-and-orange "Fire and Flames" design. The person responsible for the majority of Hendrick paint schemes (as well as firesuits, press kits, and hauler designs) is Jim Gravlin, a soft-spoken young man from Syracuse, New York. Jim, who has worked for HMS for five years, is Hendrick's art and graphics director. The computer in his cubicle displays the starting template from which he works: a single black-and-white page showing front, side, back, and top views of the car.

His palette is drawn from the DuPont paint-sample books sitting on a shelf above his desk. His choices include glittering metallic paints and high-gloss shiny paints in virtually every color. Special effects can be produced by layering paints. One page features a paint that, by itself, is a high-gloss deep blue. It's nice-looking, but not spectacular; however, layer that paint over a basecoat called "Psycho Silver," and the resulting blue sparkle almost jumps off the surface.

Three things can happen when light hits an object: The light is reflected (light bounces off the object), absorbed (light goes into the object and doesn't come back out again), or transmitted (light goes through the object). Usually, some combination of these things happens. Metals tend to reflect most of the light hitting them.

There is, however, a big difference between the gray metal of an unpainted car and a silver sparkle finish. The types of atoms from which the metal is made affect its reflectivity, but the smoothness of the finish also plays an important role. The more you polish a piece of metal, the shinier it gets. If a surface is smooth, light bounces back at an angle exactly opposite the angle it came in at. A rough surface is a bunch of small smooth surfaces oriented at different angles to each other. When the light beam comes in, it reflects not just in one direction, but in many directions (which is how wrinkle paint decreases glare).

Metallic paints, which became popular in the 1980s, are made by adding extremely small, mirrorlike metal flakes to the paint. The flakes, which are usually aluminum, are a few hundredths of a mil thick and anywhere from a few hundredths to tenths of a millimeter in width. The flake shape helps orient the reflecting surfaces approximately parallel to the car's body. Each flake has a slightly different orientation, so they catch the light from different angles and produce the twinkle effect.

Pearlescent paints have a main color enhanced by a rainbowish luster similar to that of mother-of-pearl. Pearlescent paints are part of a larger family of iridescent paints. The word "iridescence" comes

from the same root as Iris, the Greek goddess of the rainbow. Iridescent paints change colors as you look at them from different angles.

Nature has many examples of iridescence: starlings, the wings of
the blue morpho butterfly, peacock feathers, and fish scales. The first
iridescent paints used natural materials; however, it takes almost
10,000 pounds of fish to get a few pounds of scales, making that
approach a bit expensive. Mica—an iridescent mineral that naturally forms thin sheets—is used today; however, its iridescence is
very subtle. Race cars are anything but subtle.

Light waves are oscillating electric and magnetic fields. We char-
acterize the "size" of a wave by its wavelength, which is the distance
over which it repeats. Although you can't see the electric and magnetic fields, light waves behave very similarly to the wave you would
make by moving a taut jump rope up and down, for example. That
type of wave has a very large wavelength. Visible light, which makes
up a small fraction of all light waves, has much tinier wavelengths
that run from about 400 nanometers (violet) to 700 nanometers
(red), as shown in the drawing below. A nanometer is one billionth
of a meter (a meter being about thirty-nine inches). Divide the diameter of one piece of your hair by 1,000. One of those pieces
would be about 100 nanometers (nm).

Wavelength

Red light

Violet light

When two light waves are in the same place at the same time, they interfere with each other, just like two cars trying to be in the same place at the same time. How the light waves interact depends on how their peaks and valleys (the highest and lowest points of the wave) line up with each other. If the peaks and valleys of two waves with the same wavelength line up exactly, the result is a bigger wave with the same wavelength. If the two waves are exactly out-of-sync and the peaks on one wave line up with the valleys on the other, the two waves cancel each other out and you get no wave at all.

Sunlight has all of the colors in the rainbow, which means that it contains waves of many different wavelengths. The rainbow you see in a film of oil sitting on a puddle of water is created by interference between light waves. Some of the light hitting the oil surface reflects back toward your eyes. The rest of the light travels into the oil and reflects where the oil meets the water. The part of the light wave that goes through the oil layer before reflecting travels farther than the part of the wave that reflects from the top of the oil layer. If the extra distance makes the two waves exactly out-of-sync when they reach your eye (as I've drawn on the next page), you don't see that color. If the extra distance makes the two waves exactly in-sync when they reach your eye, you'll see that color more strongly.

The colors in the puddle change when you move because the distance that a light wave travels depends on the angle at which it hits the oil film. Light hitting the interface at a shallow angle travels farther than light entering at a sharper angle. Different viewing angles enhance or minimize different colors, which causes the rainbowlike iridescence.

Iridescence is possible only when you have thin layers of different materials stacked on top of each other, like an oil film on

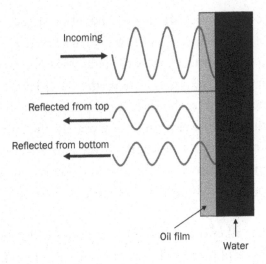

Incoming

Reflected from top

Reflected from bottom

Oil film

Water

water. Light reflects from the interfaces between those materials. The layer thicknesses must be comparable to the wavelength of visible light—which means a few hundred nanometers. Peacock feathers, pearl, and fish scales are made up of very thin layers of different materials that produce the same effect as oil on water. Natural pearl, for example, has alternating layers of calcium carbonate and protein.

Iridescent paint today contains small mica flakes coated with a very thin layer of titanium dioxide. About 50 nanometers of titanium dioxide produces a silvery iridescence. As the titanium dioxide thickness increases, the dominant iridescent color shifts: blue when the titanium dioxide is about 70 nanometers, all the way through to red at a titanium-oxide thickness of about 140 nm. However, even this better-than-nature iridescence wasn't enough for Jeff Gordon.

Among the cars displayed in the Hendrick Museum is Jeff Gordon's 1998 "Blacker," which DuPont sponsored for NASCAR's fiftieth anniversary year. "Blacker" was built in 1994 as a "top-secret" research project associated with the upcoming introduction of the

Chevy Monte Carlo. The car was named "Blacker" because it originally was painted only with black primer. The car sitting in the Hendrick Museum, however, isn't black. It's light brown. Or green. Or maybe orange.

"Blacker" goes beyond iridescence into actual color shifting. The car literally looks like it is changing colors as you move around it, thanks to a special paint called ChromaLusion (made by DuPont, of course), which takes interference one step further. ChromaLusion paint contains aluminum flakes, which reflect more light than mica. A thin spacer layer of magnesium fluoride (MgF_2) is deposited on the aluminum, and a semitransparent layer of chromium metal tops off the stack. Some light is reflected at each interface and some continues into the stack; however, the top layer is semitransparent, so when the light comes out of the MgF_2, some of it goes into the top layer and exits the flake, but some of it bounces back into the flake. A little bit of the light escapes each time, but the light reflected back creates many, many light waves that can interfere with each other. The more times the light reflects inside the flake, the more dependent the color is on where you are standing. The thinnest pigment flakes shift from gold through silver and into light blue, while the thickest pigment flakes shift from silver through green into a purplish blue.

The rich palette of colors and special-effect pigments makes just about any design possible; however, Jim Gravlin explained that there are limits on how intricate a design he can develop using paint. Color gradients take a long time to apply, as do complex shapes that have to be taped, painted one color, and then re-taped and painted another color. Until the early 1980s, everything—the sponsor logo, the driver name, and even the numbers—was hand painted.

Most inventions originate from someone getting tired of having to do the same thing over and over. In 1982, Bobby Moody, head of the body and paint shop at Richard Childress Racing, was sure there had to be a better way to get numbers on their cars than masking and painting. John McKenzie of Motorsports Designs suggested decals. Most cars today are painted with only the base color; the rest of the car is finished with custom-made decals. The decaling process takes a few hours, depending on the number and size of the decals and how many people are working on the car. The numbers on the car, the head and tail lights, the driver's name over the door, and the sponsor's logo are all decals.

The one situation in which decals pose a performance issue is on superspeedway cars. Air needs to flow around a superspeedway car as smoothly as possible, and even the raised edges of decals can provide an unacceptable increase in drag. Hendrick's Mark Thoreson explained that on superspeedway cars, they apply blank stencils before the first paint application. They apply the base color, then remove the stencils and apply the actual decals. This is followed by a clear coat applied over the entire car to produce a smooth finish.

In addition to numbers and sponsor graphics, Jim Gravlin has to leave space in his designs for a number of required decals, each of which must be displayed in NASCAR-specified positions. For example, every car has a "NASCAR Race Car" decal on the A-posts, "Goodyear" decals over the front wheels, and a host of rectangular decals on the front fenders and quarter panels that sometimes go all the way back to the door numbers.

The rectangular stickers are called contingency decals. Some companies sponsor awards for finishing first, leading at the midpoint of the race, or gaining the largest number of positions. Winning the award is contingent on your car displaying the appropriate decal in the correct location. For example, only cars displaying the "Bud" contingency sticker are eligible for the Bud Pole Award. Petty

Enterprises chooses not to display the Bud decal, so if a Petty driver wins the pole, he doesn't get the Bud Pole Award.

If you can use decals for numbers, sponsor advertisements, and the driver's name, why not just make one giant decal and do away with paint entirely? In 1997, Darrell Waltrip wanted an all-chrome car, and paint just wasn't showy enough. Motorsports Designs covered his entire car in shiny chrome-colored vinyl, and the idea of "wrapping" a car was born.

Wrapping was done at first for novelty, but Nick Woodward, Motorsports Design's sales manager, explained that wraps quickly became popular for other reasons. Teams always bring two cars to each track. The backup car—unloaded from the hauler only if the primary car is damaged beyond repair—is often a car the team plans to run at another race. For example, the backup for California is often the car the team plans to run at Las Vegas. If the team is using one paint scheme at California and a different paint scheme at Las Vegas, they would have to paint the backup car with the California design, then strip and repaint it before the Vegas race. A wrap with the California design, Nick explained, can be applied to the backup car right over the Vegas paint scheme. If the primary car survives qualifying and practice, the wrap can be removed from the backup in less than thirty minutes with no damage to the paint underneath.

This fast turnaround allows a team to run two different paint schemes in one weekend if necessary. Little Debbie's, a sponsor of the Wood Brothers No. 21 car, is a Seventh-Day Adventist company. Seventh-Day Adventists do not conduct business on Saturdays, and that includes advertising on race cars. Wood Brothers uses a wrap with an associate sponsor's paint scheme during Saturday practices and then removes the wrap to reveal the Little Debbie's design for the Sunday race.

Motorsports Designs makes wraps by printing computer-generated designs onto special vinyl developed by 3M. Overgrown ink-jet printers print on sixty-inch-wide pieces of vinyl. The printed vinyl is then covered with a clear laminate to protect the colors. A wrap is about 4 mils thick: 2 mils of vinyl and 2 mils of clear laminate, so the entire wrap is just a little thicker than a good trash bag. A wrap weighs about six pounds by itself and about nine to ten pounds once all the decals are added. Some items, like the drivers name, the car's numbers, and sponsor advertising are printed directly on the wrap, but contingency decals are often added after wrapping. It's impossible to tell whether a car is painted or wrapped when watching on television or from the stands.

Some teams, like Chip Ganassi Racing with Felix Sabates, rely almost exclusively on wraps, but wraps do have some limitations. It is difficult to print on metallic-looking vinyl. Nick Woodward showed me the sparkly silver vinyl used to wrap the Coors Light car—everything else on the car is done with decals. Printing in neon is also difficult, so if you see a number with a neon shadow, the number probably was printed—or silk-screened—on neon vinyl.

A wrap costs about $2,000, plus another $500 to $1,000 to install, which takes half a day or less. In addition to wraps, Motorsports Designs prints decals for helmets, in-car cameras, and pit boxes (some of which are wrapped as well). Although they wrapped all the cars for the movie *Talladega Nights*, Motorsports Designs deals with bigger jobs—literally—each year.

Just before the season starts, the Motorsports Designs parking lot is filled with haulers—the semi trucks that transport the cars and equipment to and from the tracks. Motorsports Designs wraps more than a hundred haulers and souvenir trailers between January 1 and the time the first hauler arrives at Daytona. Wrapping can be more economical than painting, especially for teams that lease haulers.

There is another potential benefit to wrapping instead of painting a hauler: During "Silly Season" (the time of year when drivers switch teams and sign new contracts), I'm sure some owners get a therapeutic boost from personally removing the wraps from the haulers of drivers they are happy to see leaving.

A wrap isn't actually one giant decal: There are separate pieces for the hood, roof, deck lid, rear and front bumpers, and sides. Like hanging the body, the hood, roof, deck lid, and TV panel (the vertical piece with the taillight decals) go on first, followed by the sides. The nose piece is applied last, so that it covers the edges of the side pieces and prevents them from getting torn loose.

Wraps are applied much like decals. The backing paper is peeled off and the vinyl smoothed down with a special squeegee. The primary difference between wraps and decals is their size. If you've ever hung large pieces of wallpaper, you know that getting air bubbles out from under the wallpaper is really difficult. Wraps have a grid of very tiny perforations that allow air to escape while the wrap is being applied. 3M, the company that makes the vinyl Motorsports Designs uses, has developed a new type of vinyl with air-release channels so small you can't see them. In addition to making installation even easier, the new vinyl has a glossier finish that makes it look even more like paint.

Hendrick Motorsports' Jim Gravlin doesn't look at wrapping and painting as mutually exclusive. Jim used a combination of the two techniques to make what was probably the most spectacular "paint job" of the 2006 season for Terry Labonte's last NASCAR Sprint Cup Series race before retiring. Terry had no idea when he arrived at Texas Motor Speedway that Jim had designed a very special car for the occasion. Jim's creation featured photographs from Labonte's career with Hendrick. Black-and-white images were printed on a special silver decal material. After the decals were applied to the car, three coats of DuPont's "Apple Red Candy" paint were applied on top.

Jim smiled broadly as he described the unveiling. One of the benefits of working out of a cubicle in the back of Hendrick Motorsports' administration building is that almost no one knew who he was, so he could eavesdrop on people talking about his car. When people learned he was the designer, "they kept asking if it was paint or a wrap," he said with obvious delight, "and I kept saying 'yes.'"

As I finished my tour of the Hendrick Motorsports shop and thanked my hosts, I stopped to take a last look at the shiny No. 24 car displayed in the lobby. I left with a new appreciation for not just how much craftsmanship, but also how much artistry goes into each race car.

Four

Combustion

2007 was Toyota's first year competing in the NASCAR Sprint Cup Series. Bill Davis Racing, which has been running Toyota trucks in the NASCAR Craftsman Truck Series since 2004, is Toyota's flagship team, but 2007 got off to a rough start. Dave Blaney, the only Toyota driver in the top 35 at the end of 2006 (and thus guaranteed a spot in the first five races of 2007) had promising runs ended prematurely by bad luck. I visited BDR the week after the second race of the season at Fontana, where the No. 22 Caterpillar Toyota qualified 14th and was having a good run until its engine gave out on lap 113.

Bill Davis Racing is located about eighty-five miles north of Charlotte in High Point, which is the only city in North Carolina that spans four different counties. The shop is located in a more industrial setting than the business parks of Concord, consistent with the city's strengths in furniture, textiles, and bus manufacturing. As I entered the brown brick main building, the official BDR greeter, Vegas, trotted out from behind the receptionist's desk to sniff my hand and politely ask for a scruff behind the ears. The friendly Briard belongs to owners Bill and Gail Davis. The receptionist directed me across the street from the main shop to the BDR engine research building.

Andy Randolph, a lanky, enthusiastic man with graying hair, is BDR's engine technical director. He clearly loves what he does, as evidenced by how often he punctuates sentences with laughter. Andy worked on GM's diesel engine program after receiving his Ph.D. in chemical engineering from Northwestern University in 1985. He never expected to end up in NASCAR.

"The GM race shop called me one day out of the blue to ask if I would help an intern doing some combustion studies on their race engines," Andy explained. "So I went over to McLaren Engines up in Detroit with this intern and started taking some data and it was very obvious where they could gain performance fairly easily. I told the fellow right there what we needed to do to improve performance and he looked at me like I was crazy, but he pacified me and did it and we made a significant gain in engine output right there. So all of a sudden, they wanted me to not only teach the intern, but do combustion everywhere. I went down to North Carolina and would help on my vacations." This experience led to consulting for Richard Childress Racing and Robert Yates Racing, the latter during Dale Jarrett's 1999 NASCAR Sprint Cup championship season.

Andy was just moving into a new office when I visited. He came to BDR after seven years at Hendrick Motorsports. "I just feel like I was in the right place at the right time," Andy said. "I think we have a tremendous amount of potential for the future."

Andy's Ph.D. thesis was on combustion, the process that converts the energy stored in gasoline into motion—and, in the case of a stock car, lots of it. Gasoline molecules belong to the hydrocarbon family, which means they contain hydrogen and carbon atoms. Gasoline, kerosene, natural gas, and paraffin are all hydrocarbons, with the main differences between family members being how many carbon atoms each molecule contains and how the atoms are arranged.

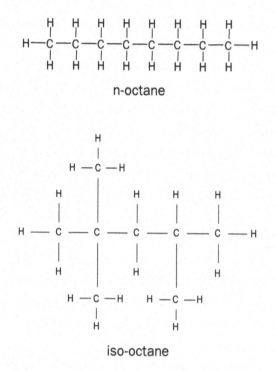

n-octane

iso-octane

The octane molecule (*octa* means eight) has the formula C_8H_{18}, which means that there are eight carbon atoms and eighteen hydrogen atoms; however, the formula doesn't tell the whole story because the atoms can be connected in different ways, as I've drawn above. The molecule on the top is called n-octane ("n" for "normal") because the eight carbon atoms are in a straight line. In contrast, the eight carbon atoms in iso-octane have a branched arrangement. Some hydrocarbon molecules even have rings of carbon atoms. Gasoline is a mix of hydrocarbons with most molecules having between five and twelve carbon atoms.

Gasoline molecules—like all molecules—store energy in their chemical bonds. When gasoline reacts with oxygen, the bonds holding atoms together in the gasoline and oxygen molecules break, and those atoms can then make new bonds with atoms they like better than their former partners. The carbon atoms from the gasoline bond with oxygen atoms to make carbon dioxide and the hydrogen atoms bond with oxygen to make water. The products (water and carbon dioxide molecules) store less energy in their chemical bonds than the reactants (gasoline and oxygen molecules). The energy left over is what makes the car move. Cellular respiration, which is how your body extracts energy from food, is also combustion, but you use sugars (molecules made of carbon, hydrogen, and oxygen atoms) as fuel instead of hydrocarbons.

The combustion equation for octane is: $2C_8H_{18} + 25O_2 \rightarrow 16CO_2 + 18H_2O + Energy$, which tells you that 2 octane molecules (C_8H_{18}) combine with 25 oxygen molecules (O_2), producing 16 molecules of carbon dioxide (CO_2), 18 molecules of water vapor (H_2O), and—the important part—energy.

It takes a surprisingly large number of parts to turn that energy into speed, as Andy showed me during a tour of the engine shop. The remains of Dave Blaney's Fontana engine were laid out neatly on the shelves of rolling carts, almost as if for a funeral viewing. Small pools of oil had collected beneath the parts, contained by lips on the shelves.

Although this engine was—well, make that had been—much more powerful than a production-car engine, the majority of the parts and the way the engines work are very similar. I've sketched some of the most important parts on the next page, which shows one combustion chamber from an engine.

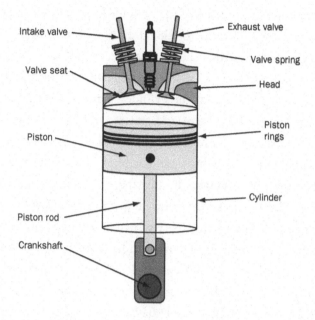

The combustion chamber is where the chemical energy stored in gasoline is converted into motion energy. A piston, which is sealed to the edges of the cylinder by its piston rings, moves up and down. The crankshaft changes the up-and-down motion of the piston into

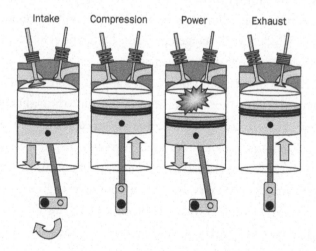

rotational motion, much like the up-and-down motion of your legs on a bicycle turns into the rotational motion of the wheels.

Most car engines use the four-stroke cycle I've drawn on the previous page. The cycle is continuous, but we'll pick it up when the piston is at the top of the cylinder. During the intake stroke, the piston moves downward and creates a partial vacuum. The intake valve opens to let a mixture of fuel and air into the cylinder. In the second step (the compression stroke), the intake valve closes and the piston moves upward, squeezing the fuel/air mixture into a much smaller space. In the third step (the power stroke), a spark from the spark plug (which isn't shown in the drawing) starts the combustion reaction, producing a sudden expansion of hot combustion gases. Both valves are now closed, so the force of the combustion gases pushes the piston downward. In the fourth and final step (the exhaust stroke), the exhaust valve opens and the piston moves up to push the combustion gases out of the cylinder. These four steps repeat over and over.

Internal combustion means that combustion occurs in a confined space that, in this case, is formed by the cylinder and the head. The cylinder is literally a cylindrical hole in the engine block. A cylinder produces power only during its power stroke, so most engines have multiple cylinders with staggered cycles. Different cylinders execute their power strokes at different times.

The engine block from the No. 22 car was mounted on a stand that allowed Andy to rotate it easily and show me the eight cylinders arranged in a V-shape with four cylinders on each side. None of the NASCAR-approved blocks are used currently in production vehicles, although their close relatives do appear in consumer cars. The rotating stand is necessary because a complete engine can weigh more than 500 pounds, with the block accounting for about 180 pounds of that weight. Engine blocks can be made from aluminum alloys, but those alloys are expensive, so NASCAR limits teams to cast iron.

Engine blocks are made using a process called sand casting. A design is pressed into wet sand and used as a mold. The molten metal solidifies in the mold, and then the mold is removed (or broken) to extract the piece. The process sounds crude, but special sands and binders make casting an excellent technique for fabricating intricate parts like engine blocks.

Cast iron has between two and four percent carbon by weight, which makes it stronger than the mild steel used in the chassis. Like mild steel, the specific structure of cast iron is determined by the other elements added and how the molten metal is cooled. Iron atoms, you may remember, are happy to have carbon atoms around when the mixture is liquid; however, iron doesn't like carbon getting in the way when it's time to solidify. Iron atoms will grudgingly form cementite if they must, but would much prefer the carbon atoms to go off and solidify by themselves.

The two types of regular cast iron are named according to their appearance when broken. White cast iron has a whitish crystalline surface when fractured and is made by cooling the iron-carbon liquid quickly enough to form cementite grains. The carbon doesn't have time to get out of the way before the liquid becomes solid. White cast iron is very hard, but also extremely brittle and difficult to machine. Gray cast iron, in contrast, has two to three percent silicon (by weight) and a gray surface when broken. Gray cast iron is formed by cooling the molten alloy slowly enough to allow the carbon time to solidify into small flakes of graphite.

Graphite is a form of carbon in which atoms are connected strongly within planes, but the planes are weakly connected to each other. The graphite flakes in gray cast iron are long and thin, with sharp edges. Their lengths range from the diameter of a human hair to about one-tenth that size, depending on the particular recipe used. Gray cast iron is fairly strong and easier to machine because the graphite flakes serve as lubricants.

But gray cast iron is still pretty brittle, which means it cracks easily. The sharp edges of the graphite flakes concentrate stress, just like a high-heeled shoe makes your foot hurt more than a flat shoe. A high heel concentrates your weight, while a flat shoe distributes your weight over a larger area. The graphite flakes concentrate stress the same way, which means that once a crack starts, it can move easily through the entire block.

Cast iron would be a better material if you could keep the graphite but lose the sharp edges. In the late 1940s, metallurgists—people who study metals—figured out that adding small amounts of magnesium (as little as a few hundredths of a percent by weight) changes the graphite shape from flakes to rounder structures called nodules that don't concentrate stress the way flakes do. The resulting ductile cast iron (also called nodular cast iron) is about twenty times more ductile than gray cast iron.

The problem is that ductile iron is a little *too* ductile. Although it is more crack resistant and stronger than gray iron, ductile cast iron is harder to cast into parts and it isn't as good at dissipating heat as gray cast iron.

An extremely narrow range of magnesium concentration produces a compromise: wormlike graphite structures that look like pieces of a miniature coral reef. The randomly oriented pieces interlock within the iron matrix. This compacted graphite iron, or CGI, is one and a half to two times stronger than gray cast iron, and dissipates heat better than ductile iron. CGI is also twice as resistant to metal fatigue (weakening due to repeated cycles of heat and stress) as gray cast iron.

CGI was discovered about the same time as ductile cast iron, but it hasn't been used in production until recently because the magnesium concentration has to be controlled to within a few thousandths of a percent. A one-pound addition of magnesium has to be weighed with an accuracy of thousandths of an ounce. The only way to achieve this level of accuracy is with computer-controlled manufacturing processes.

All NASCAR engines have CGI blocks. CGI's higher strength means that parts can be made with thinner walls, which produces finer details and lighter parts. The Toyota Racing Development truck engine has cylinder walls only twelve hundredths of an inch thick. CGI engine blocks are starting to appear in production vehicles, where they can make cars more efficient by reducing engine weight and size. Tolerances of tens of thousandths of an inch or less are possible; however, Andy explained that he isn't necessarily concerned with getting high-precision castings.

"For the cylinder head, for example, my preference would be a block of aluminum that's this big and this tall and this wide," Andy laughed as he outlined a block about two-and-a-half feet on a side with his hands, "so that we can cut in the ports, we can cut in the chambers, we can put anything where we want it."

Andy doesn't have quite that much freedom in designing his engines. NASCAR requires teams to use blocks, cylinder heads, and intake manifolds from NASCAR-approved manufacturer castings. NASCAR also limits the modifications Andy can make to those parts. He cannot change the number of cylinders, the bore angles, the separation between the bores, or the location of the camshaft in the block, for example. He can, however, make the cylinders larger (up to the NASCAR limit of 4.185 inches) and rounder, install thin-wall press-in cylinder liners, and apply surface treatments to improve the frictional, thermal, or wear properties of his engines.

Even if NASCAR allowed him more flexibility, Andy would

still be constrained by the extreme conditions under which the engine works. The cylinder and all the parts in it normally operate at temperatures up to 2,000°F (1,093°C) and pressures up to 1,500 pounds per square inch (psi). 1,500 psi is equivalent to a thousand-pound steer putting all of its weight on a quarter. These are extreme conditions—and that's when everything is working correctly.

Engine trouble is the second-most common cause of a car not finishing a race. (Accidents are the first.) The phrases used to describe engine problems are violent: "blowing an engine," "throwing a rod," or "melting down." Looking at what is left of the Fontana engine, I can assure you that these are not just metaphors. Some of the bearings on which the crankshaft rotates have literally melted. A couple of the pistons are singed. Although the parts are designed to sustain the high pressures and temperatures of combustion, sometimes the combustion process doesn't go as planned.

When the spark plug sparks, the first combustion reactions should produce a small flame kernel about the size of the spark-plug gap. That kernel grows and expands more rapidly through the cylinder as it consumes the rest of the air/fuel mixture. The flame front (the area actively combusting) moves at roughly 300 feet (91 meters) each second.

If the pressure or temperature gets too high, all of the fuel/air mixture can combust at once. The resulting explosion is called autoignition and produces a flame front that can move more than a hundred times faster than the controlled burn of combustion. The gases produced create a shockwave, which is a sharp boundary between moving gas and still gas. That shockwave bouncing around the cylinder produces the "knocking" sound that alerts you to possible engine trouble.

If the cylinder gets hot enough, the fuel/air mixture can ignite *before* the spark plug fires, which is called preignition. Both processes can wreak havoc in the engine by producing extreme pressures—up to and even greater than 3,000 psi. If the piston hasn't finished traveling upward when autoignition occurs, the force of the explosion can push the piston back down hard enough to bend the rods connecting the pistons to the crankshaft. The temperature in the cylinder can escalate rapidly. I asked Andy if pistons actually melt.

"Oh, yeah!" he laughed. "You take something going at 800 horsepower and break it and lots of damage happens." Since he'd just moved, he hadn't built much of a destroyed-parts collection. He apologized for not having any melted pistons available to show me.

Even without mishaps, the heat produced by combustion needs to be removed from the engine. Heat can only be transferred between areas with different temperatures, and can only move from hot to cold. If you put a metal spoon in a cup of hot coffee, the atoms at the end of the spoon get warm and start to move more quickly. Those atoms start the atoms next to them moving, which start the atoms next to them moving, and eventually the spoon handle gets warm. This won't happen if you use a plastic or wooden stirrer. Plastic and wood aren't very good at moving heat from one place to another.

Heat in solids is due to atomic motion, but understanding why metals are good heat conductors but plastic and wood are not requires looking a little more deeply into the atom. An atom is made of a nucleus and electrons. The nucleus is large and heavy, while electrons are smaller and much lighter. Although electrons aren't as physically impressive as the protons and neutrons in the nucleus, electrons can do things the nucleus cannot.

Nuclei in solids have assigned positions. They can vibrate about their assigned positions, but they can't move very far. They can, however, allow some of their electrons to wander away, and those moving electrons transfer heat. Thermal insulators and conductors both transmit heat via vibrating nuclei; however, conductors can transmit heat by electron motion. Different metals are more or less lenient about how many electrons they let wander and how far they can go, so some metals are better thermal conductors than others.

Thermal insulators are more selfish with their electrons, so electrons in these materials don't play much of a role in transferring heat. In addition to keeping their electrons close, materials like wood, plastic, and rubber tend to have less orderly atomic arrangements than crystalline metals. Ordered atomic arrangements transfer heat more effectively, which gives metals another advantage in moving heat away from the engine.

When the surface of a metal gets warm, the atoms start to move more quickly. Those atoms transfer heat energy to atoms farther away from the surface. As the heat continues to move through the material, more and more atoms share the heat energy, and the overall temperature decreases. If the atoms at the surface can't get rid of heat fast enough, the temperature rises and the metal can melt.

Heat transfer is one reason most engine parts are made of metal; however, good heat transfer isn't the only requirement. A piston has to be strong enough to withstand high pressures, and it has to move quickly. A piston in a 9,500-rpm engine moves up and down at almost one hundred feet per second. Steel is strong and dissipates heat well, but a steel piston has a lot of inertia. Inertia is how strongly something resists moving (or stopping if it's already moving). Mass is a measure of inertia. The more massive something is, the harder it is to move.

Steel is a dense material. The atoms in steel are heavy and packed closely together. You could decrease a steel piston's inertia by making it smaller, but a smaller piston isn't as strong and doesn't transfer heat as well. Aluminum mixed with copper, iron, magnesium, and nickel (plus small amounts of a few other elements) has the required strength and thermal conductivity and is about a third as heavy as a similarly sized piece of stainless steel, which is why the pistons in a race engine are made from aluminum alloys.

Even with very good thermal conductors, the engine needs help getting rid of heat. Water is pumped in a closed cycle through passageways in the engine block and head, and then through the radiator. The water entering the engine is cooler than the engine block, so molecules in the engine block transfer heat to the water molecules racing past. When the now-warm water molecules reach the radiator, they transfer heat to atoms in the many small metal fins that stick out from the radiator into the cooler air. Heat goes from water molecule to metal atom and from metal atom to air molecule. Only metal atoms in contact with the air can transfer heat. The larger surface area of the fins puts more metal atoms in contact with the air. If the fins are damaged in a crash—even if the radiator isn't ruptured—the loss of surface area can slow heat transfer enough for the engine to overheat. After dumping their heat in the radiator, the now cooler water molecules reenter the engine block to pick up some more heat. It's like a shuttle service: The cooling water takes heat from the engine block to the radiator, and then goes back to the engine for more.

Drivers routinely report water temperatures in the 220° to 250°F range, which is well above the 212°F (100°C) boiling temperature of water. Engine designers make this possible by placing the cooling system under pressure. Water boils at 212°F, but only when the water is at the atmospheric pressure of 14.7 psi. When the pressure is lower—as it is at high altitudes—water boils at temperatures below 212°F, and when the pressure is higher, water won't boil until it gets hotter than 212°F.

Molecules in water are constantly moving, not only within the water, but also to and from the air surrounding the water. Some water molecules have enough energy to escape the liquid and become water vapor. Some water vapor molecules lose energy and return to the liquid. Boiling means that enough water molecules have moved from the liquid to the vapor to make the pressure of the water vapor the same as the pressure of the surrounding air. If the air is at the atmospheric pressure of 14.7 psi, this occurs at a temperature of 212°F. If the air is at a higher pressure, more energy—meaning higher temperature—is needed before the water will boil.

Most teams use a 30 psi radiator cap, which allows the pressure inside the cooling system to reach a maximum of 30 psi *above* atmospheric pressure. Water won't boil until around 275°F (135°C) at this pressure. The higher pressure keeps the water from turning into steam until it reaches higher temperatures; however, warmer water doesn't cool the engine as efficiently as cooler water.

If the water in the radiator does get above its boiling point, the pressure will increase as water changes from liquid to steam. As soon as the radiator pressure gets above the radiator cap's pressure limit, steam (or very hot water) will spurt from an area near the cowl on the right side of the car. This is a signal to the driver (and everyone else) that the engine is starting to overheat. Expelling some water or steam gets the pressure back down below the 30 psi limit. (This is also why you should never remove the radiator cap from a hot car: The only place for the hot water and steam to escape from your car is through the radiator cap when you remove it.)

Heat is a form of energy, which means that all the heat being expelled by the radiator is lost energy. You've heard that energy is never created or destroyed, just changed from one type of energy to another. When we talk about energy being "lost," we mean that it's been converted to a form we can't harness for useful work. When energy changes into heat, light, or sound, for example, most cars have no way to recapture that energy and use it to make the car move. The energy is "lost" in the sense of not producing power, but not lost in the sense of having disappeared.

Most gasoline-fueled internal combustion engines "lose" about 75 percent of the energy in the gasoline to heat, which is either removed by the cooling system or expelled as exhaust gases. Some energy is used to run the water pump and the air conditioner, and the rest of the energy goes to overcoming friction.

There are some places we want friction. Air resistance is friction between air molecules and the car, which can produce drag (bad) and downforce (good). The friction between the tires and the track is necessary for the car to move. The engine, however, is one place where we really want to minimize friction. Friction comes from things rubbing together, and rougher surfaces produce more friction.

A sheet of 60-grit sandpaper is bumpy. You won't see any bumps on a sheet of aluminum foil—or the sides of a piston ring—but that's only because you aren't looking closely enough. If you could see the seemingly smooth surface of a piston ring at a high enough magnification, you would find lots of bumps, protrusions, and dips—just on a smaller scale. If you look closely enough, the surface of any material looks like sandpaper. Friction is these bumps rubbing across each other. The amount of friction depends on how the bumps interact with each other when they make contact. Bumpier materials make more friction, and pressing the two pieces together harder also creates more friction.

Friction can be problematic in multiple ways. If the two materials have different hardnesses, one can wear down the other (like sandpaper on wood). Friction also creates heat. Rub your hands together for a few moments and then put them to your face. Some of the motion energy from your hands has been converted to heat. If there's one thing we don't need more of in the engine, it's heat.

You can avoid friction by not allowing things to rub against each other; however, the only way to harness the energy of combustion is to seal the gases in the combustion chamber. If they leak out, they can't push the piston down. We need to seal in the gases while simultaneously minimizing friction.

The solution is oil. Engine oil seals, lubricates, cools, cleans, and prevents rust. Oil is a liquid, which is one of three (four if you include plasma) fundamental phases of matter. The molecules in liquids aren't attached to each other as rigidly as the atoms in a solid, which means that liquid molecules can slide past each other more easily than the atoms in solids can.

Oil forms a thin film between two parts. The piston rings (usually two or three of them) fit into a series of grooves around the piston's circumference. The top rings make the seal and the bottom ring supplies oil. Oil molecules get between the metal pieces and prevent gas molecules from escaping, even when the piston is moving. They also allow the piston to move in the cylinder without the two pieces of metal actually touching. Every moving part in the engine is lubricated. The crankshaft bearings, connecting rods, camshafts, and piston rings all rely on oil films to move smoothly.

The problem with oil is that it can make the engine work harder. Water rushes out of a jug, while syrup oozes. The "flow-ability" of a liquid is measured by its viscosity. The higher a liquid's viscosity, or "weight," the harder it is to pour. Oil viscosity is determined by how the oil molecules interact with each other.

Watch Dale Earnhardt, Jr. try to walk through the garage: He moves slowly because everyone wants an autograph. You or I, on the other hand, could get across the garage much faster because no one is interested in our autographs, so we don't interact with as many people or for as long a time. Liquids behave similarly. When two molecules get close to each other, both slow down. How much they slow down depends on how fast they were going before they met and how much the two molecules like interacting.

An oil's "weight" doesn't have anything to do with its actual mass. Oils commonly come in weights from 0 to 50, with lower-weight oils being less viscous than higher-weight oils. A zero-weight oil looks and pours like water. More viscous oils make thicker films, but also offer more resistance to parts moving past each other. It is much easier to swim in water than it is to swim in pancake syrup.

Heating syrup makes it much easier to pour because the syrup becomes less viscous. Oil works the same way: The higher the temperature, the less viscous the oil becomes. While this is good for syrup, it's not always good for motor oil. Oil has to have enough viscosity to form and maintain a film between parts. The oil in a cold engine is thick and may not be able to get between the moving metal parts fast enough when the engine first starts. When oil gets too hot, the pressure produced by two pieces moving past each other can break the oil film and allow the metal pieces to touch. Racing engines use lighter-weight oils because the clearances between parts are smaller than in production engines. High-viscosity oil has a harder time getting into small spaces.

Lighter-weight oils offer less resistance to motion, so teams may use very lightweight oils for qualifying to eke out a few extra horsepower. Andy explained that many teams use as low as zero-weight oil for qualifying at superspeedways, 10- or 20-weight oil for superspeedway races, and 30-weight oil for everything else. A zero-weight oil would become too thin after heating up over thirty or forty green-flag laps, but it works just fine for the two laps necessary to get a good starting position at Daytona.

Most of the oil you buy has two weight numbers, like 5W30. The first number indicates the oil's viscosity at lower temperatures. The second number tells you that the oil doesn't thin more than an oil of that weight would at higher temperatures. A 5W30 oil behaves like cold 5-weight oil when cold but like warm 30-weight oil when warm. Regular oil (in which about 90 percent of the molecules are hydrocarbons, with sixteen to fifty carbon atoms per molecule) doesn't do this on its own.

Getting a substantially different oil behavior at high temperatures requires adding polymers that change shape when they get warm. At low temperatures, the polymers coil up into little balls and don't really affect the oil's viscosity. When the oil warms up, the polymers uncoil into long chains. The polymers increase the oil's viscosity, but only when it becomes warm enough for them to uncoil. Going back to our example of Dale Earnhardt, Jr. trying to cross the garage, these unfurling polymers would be like placing haulers in his path: He'd have to travel around the haulers instead of going straight. This would further extend his trip and he'd run into more people.

In addition to lubricating, oil also helps cool the engine. Andy explained that the valve springs move up and down rapidly, which makes them extremely hot. Oil-squirter towers are mounted in the cylinder heads and squirt oil on the valve springs to cool them. Oil squirters also are used underneath the pistons to manage heat there.

The large heat load requires a lot of oil. The four or five quarts of oil your car uses are stored in an oil pan (or sump) beneath the crankshaft. NASCAR engines use a 22-quart dry sump, which means that the oil is stored in an oil tank and not at the bottom of the engine. The dry-sump oil tank sits behind the driver. There is no oil pan to get punctured or bang on the track, so the engine can be mounted lower in the car.

Oil helps clean the engine by picking up small particles like those produced by wear between engine parts and depositing them in the oil filter. These particles, along with sludge from unburned fuel and soot, are what makes the oil you drain from your car after 5,000 miles much darker than it was when it went in. Sludge, by the way, has the same effect on an engine that plaque has in your arteries. Sludge can narrow the passageways the oil has to get through, which decreases the oil's flow rate. Sludge isn't a problem for NAS-CAR teams because the engines are torn apart completely after every use.

In addition to everything else it does, oil keeps the people who make shop wipes (an industrial-strength paper towel product) in business. As Andy wiped some oil from his hands, our conversation was interrupted by a low growl. The growl accelerated into a whine, which continued for about ten seconds, and then stopped. I looked questioningly at Andy.

"Dyno test," he said, suggesting that we return to his office for a brief discussion about what exactly a dyno tests—and what a dyno is—before going to take a look at what's making all that noise.

Five

Power Play

Andy Randolph, the engine technical director at Bill Davis Racing, has energy to spare. If he needs another nickname (everyone in NASCAR seems to have at least one), I suggest "Tigger," because Andy has the same bounding enthusiasm as the Winnie the Pooh character. I asked if he watches races from the pit box, which is the small raised structure where the crew chief and perhaps four other people can sit to get a better view of the track. He responded that he watches from nearby because "It's hard to pace in a pit box."

The racing business is an ideal place for someone with Andy's energy. "If you have a good idea at GM," he said, "it might take you three or four years to see it in a car, whereas here, it can be three or four days." An academic career would probably be too sedate for him, but he is a natural teacher. In addition to his very clear explanations, he can't resist using each of the different colored dry-erase markers sitting on the whiteboard in his office.

"Let me draw you a picture here," he said. With a red marker, he drew the standard diagram you'd find in any physics textbook showing how the pressure and volume change during a four-stroke engine cycle. The diagram has nice straight lines with a broad parallelogram in the middle.

"Of course," he laughed, "that's not reality." In blue, he drew a second diagram over the first, with wavy lines over the straight and, instead of the neat parallelogram, a slanted oval. Andy shaded the areas between the red parallelogram and the blue oblong shape inside it. The shaded area is all lost work. "You have exhaust loss, you have intake loss, you have heat loss," he rattled off. "Combustion doesn't happen instantaneously, and the exhaust valve doesn't open instantaneously."

Testing tells Andy how close their engines are to the ideal red graph on the board. A dynamometer—dyno for short—is like a treadmill test for an engine. BDR's dynamometer is housed in a square room about twelve feet on a side. A four-by-three-foot window with bulletproof glass lets the engineers outside watch what's happening. Andy showed me a video of an engine literally exploding during a dyno test, which confirms the need for bulletproof glass.

The engine being tested looks like an intensive-care patient, with hoses and wires everywhere. A piece of corrugated tubing, like the kind you use in dryer vents, but larger, runs to the engine's air intake. Another piece of tubing carries off exhaust. Bundles of cables and wires run from inside the dynamometer room to computers outside. Sensors measure everything from the chemical composition of the exhaust gases to the pressure and volume in each cylinder. The output from the latter two sensors is displayed on the screen and—surprise— looks just like the blue oval Andy drew on his office whiteboard.

A dyno doesn't measure power directly. A dynamometer measures torque. Torque is twisting ability, like you use for turning bolts or crankshafts. The torque is the force you apply times the perpendicular distance between where you apply the force and the center of rotation. If you're having a hard time loosening a lug nut, try using a longer wrench. If you exert 30 pounds of force on a foot-long wrench, you've exerted 30 foot-pounds of torque. If you use a two-foot-long wrench, the same 30 pounds of force gives you 60 foot-pounds of torque.

Torque determines how fast the car can accelerate, and the amount of torque is determined by the force with which the pistons rotate the crankshaft. An engine's torque is always measured at a specific engine speed. The engine speed is how many revolutions per minute (rpm) the crankshaft makes, and the engine produces different amounts of torque at different engine speeds. The dynamometer provides varying amounts of resistance to rotation, which changes the engine speed. The amount of torque the engine produces at each speed is recorded and looks like the solid line on the graph below.

The engine's horsepower is calculated by multiplying the torque (in foot-pounds) by the rotational speed of the engine (in rpm) and dividing by 5,252. The unit of horsepower was introduced by James Watt, the inventor of the steam engine, and appears to have been more of a marketing tool than a true measure of a horse's power output. I've shown the horsepower as the dashed line on the graph. The metric unit for power is the watt, so engine power may be given in kilowatts, which are thousands of watts.

You usually hear only about an engine's maximum torque and horsepower, but the shapes of these curves are important. Toyota made a mid-course correction just before the 2007 Coca-Cola 600 when they realized that although their engines had just as much

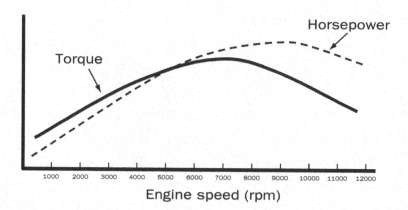

peak horsepower as anyone else, the torque curve didn't rise as rapidly. Less torque means less ability to accelerate, so the Toyota drivers couldn't get on the throttle coming out of the corners as quickly. After the changes, five of the seven Toyotas qualified for the Coca-Cola 600, and Brian Vickers became the first Toyota driver to earn a top-5 finish in a Cup race.

Andy can change the shapes of the torque and horsepower curves by changing the engine structure. The torque and power depend on the dynamics of the combustion event and how fast gas gets into and out of the cylinder, both of which are controlled by the camshaft. Your car probably has overhead camshafts, which means that the camshafts are located above the valves. The single camshaft in a NASCAR engine is mounted in the block, above the crankshaft, as I've shown below. The camshaft, which is connected to the crankshaft by a belt and rotates at half the speed of the crankshaft, has two lobes per cylinder, one lobe for each valve. As the camshaft rotates, each lobe pushes a valve lifter, which raises a pushrod that lifts one end of the rocker arm. The other end of the rocker arm is connected to the valve. When the pushrod moves up, the rocker arm opens the valve.

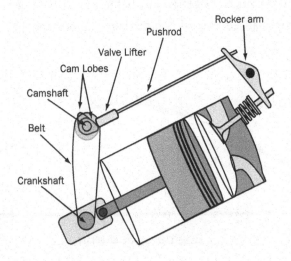

The lobe shape determines when the valves open, how quickly they open, and how long they stay open. Andy explained that the cam profiles on NASCAR engines are very different from those on production car engines. The NASCAR engine intake valves open earlier, open about twice as wide as those on a production car, and close later, which helps the fuel/air mixture move into the cylinder faster and the exhaust gases get out faster.

The engine dyno gives Andy a report card of how changes they make to the engine affect its torque and power output. One of BDR's engines produces almost three and a half times the power of a stock V6 Camry engine, as I've shown in the table below. Power is the rate at which we use (or produce) energy. People consume energy the same way cars do: The engine gets its energy from gasoline and spark plugs, while we get our energy from food.

	2006 Toyota Camry LE V6	NASCAR Engine
Number of cylinders	6	8
Bore×stroke (in)	3.70×3.27	4.185×3.25
Camshaft type	Dual Overhead	Pushrod
Displacement per cylinder (in³)	35.2	44.98
Total displacement (in³/L)	211/3.46	358/5.867
Compression ratio	10.8:1	12:1
Number of valves per cylinder	4	2
Torque (ft-lbs)@rpm	248 @ 4700	550 @ 7500
Horsepower (hp)@rpm	268 @ 6200	850 @ 9200

Two people can expend the same amount of energy but generate different amounts of power. You and I can both burn a hundred calories of energy by exercising, but if you run and do it in twenty minutes, and I take a one-hour walk, you've generated more power. You expended the same amount of energy as I did, but you took less time to do it.

The average person can produce a sustained power output of about a tenth of a horsepower, which is roughly the power used by a 100-watt lightbulb. Trained athletes can manage up to about a third of a horsepower for a period of several hours, and a highly trained athlete, like a world-class cyclist, can output 1½ horsepower for very short periods of time.

Most NASCAR engines have a maximum power output of around 850 horsepower, which is equivalent to 151 food calories per second. One plain M&M has about 4.4 food calories. You would have to eat thirty-four plain M&Ms *every second* to match the power of a NASCAR engine. The BDR engine has about three-and-a-half times the power of the street Camry engine, which means that if both engines run for ten minutes, the NASCAR engine produces three-and-a-half times as much energy as the street engine. Is the difference due solely to size?

In the 1950s and '60s, engine size was the primary selling point for cars. Bigger was definitely better. The Beach Boys' song *409* refers to an engine displacement (measured in cubic inches), not to a model number. Manufacturers scaled back engines when gas became scarce in the 1970s, but that created a marketing problem. The 1971 Chevy Vega engine was 140 cubic inches, which looked simply pathetic next to the 500-cubic-inch engine of a 1970 Cadillac. The metric system came to the rescue: GM advertised the Vega as having a 2300 cc (cubic centimeter) engine, which at least made it *sound* big.

The engine displacement (how we measure engine size) is the displacement of one cylinder multiplied by the number of cylinders. The displacement of a single cylinder is the cylinder area times the vertical distance the piston travels, which is called the stroke. The NASCAR engine has larger cylinders than the street Camry—and more of them. The 3.5-liter V6 street Camry engine has a cylinder diameter (called the bore) of 3.70 inches and a stroke of 3.27 inches, which makes the cylinder's displacement about 35.2 cubic inches. The total engine displacement of the Camry is 211 cubic inches, while the NASCAR engine displacement is required to be between 350 and 358 cubic inches. Although the street engine has about 60 percent of the displacement of the NASCAR version, it generates only about a third of the horsepower.

A larger cylinder produces a number of advantages. It holds more fuel/air mixture, so you get more energy per power stroke; however, a large cylinder bore also gives you room for larger intake and exhaust valves. Larger valves move the fuel/air mixture in and the exhaust gases out more quickly.

Engine torque is related directly to the force exerted by the piston. That force is the product of the pressure created by the combustion gases times the area of the cylinder. The area of a street Camry cylinder is 10.75 in². The area of a NASCAR cylinder with a bore of 4.185 inches is 13.76 in². If both cylinders experience the same pressure, a street Camry engine produces only 78 percent of the force produced by the NASCAR engine because its piston area is smaller.

That's if the pressure in the two cylinders is the same—but it's not. The higher the pressure in the cylinder before combustion, the higher it will be after combustion. The pressure before combustion is determined by the compression ratio, which is the ratio of the maximum volume in the cylinder (when the piston is at its lowest position) to the minimum volume in the cylinder (when the piston is at its highest position).

NASCAR limits compression ratios to a maximum of 12:1 (twelve to one), which means that the volume of the fuel/air mixture has been compressed to one-twelfth the starting volume during the compression stroke. That's not so different from the 10.8:1 compression ratio of a street Camry—it gives the NASCAR engine a few percent more power. The higher horsepower of the NASCAR engine can't be accounted for by only its larger size.

What about the fuel? NASCAR engines use higher octane gas than production car engines. When you fill up your tank, you usually have a choice of three gasoline grades: Regular gasoline (usually 87 octane), 10 percent ethanol (usually 89 octane) and premium, which usually is between 91 and 93 octane.

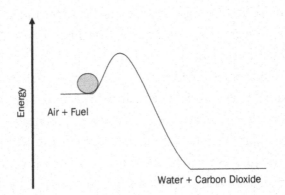

A fuel's octane rating doesn't tell you how much energy it contains— it tells you how easy the fuel is to ignite. Gasoline and oxygen don't spontaneously combust when mixed together: They need a spark from a spark plug to start the combustion reaction. I've illustrated this by analogy in the picture above. The higher the ball is, the more energy it has. If you can provide just enough energy to get the ball to the top of the hill, it will roll by itself to the bottom of the other side, where it has much lower energy. The difference in energy between the left side of the hill and the right side of the hill is the energy that can be used by the car. You have to add a little energy (the spark, which originates at the battery) to get a bigger amount of energy out.

How much energy you have to add is different for different molecules. Each type of gasoline molecule has its own autoignition temperature—the lowest temperature at which the molecule combusts without adding any additional energy. The n-heptane molecule (another hydrocarbon) autoignites at 432°F (222°C), while the iso-octane molecule won't autoignite until 880°F (471°C).

The autoignition temperature decreases as the pressure in the cylinder increases. The molecules in diesel fuel will start combustion just by being compressed enough. Gasoline doesn't combust as easily as diesel fuel, so even though the pressure in the cylinder increases quite a bit, we still need a spark plug to provide the rest of the energy necessary to start the combustion reaction.

Octane numbers were assigned arbitrarily: n-heptane was assigned a value of 0 and iso-octane a value of 100. Gasoline with an 87 octane rating has the same resistance to autoignition as a mixture of 87 percent iso-octane and 13 percent n-heptane by volume. To make matters more confusing, there are different kinds of octane ratings: The research octane number (RON) is higher and reflects the "best case" conditions, while the motor octane number (MON) is measured under more stringent conditions and is lower. In the United States, the number on the gas pump is the anti-knock index, which is the average of the RON and the MON. (This is what the formula (R+M)/2 on the pump means.) Unless otherwise indicated, I'll refer to fuels by their anti-knock numbers.

A higher octane rating means that the fuel/air mixture can withstand higher pressures before autoigniting. Higher cylinder pressures, like those produced by large compression ratios, require higher octane fuel. Using a higher-octane-rated fuel than specified for your engine will not improve your car's performance and may actually make it worse. Saying that higher octane gas is better is like saying that a recipe calling for a teaspoon of salt would be better if you used a tablespoon of salt.

Until the 2007 season, NASCAR used fuel with an anti-knock index of 112. If octane ratings go from 0 to 100, how do you get 112-octane gasoline? The octane scale is arbitrary in the same way that temperature scales are arbitrary: The temperatures at which water freezes and boils were chosen as the two fixed points on the temperature scale because they were easy to measure. Similarly, n-heptane and iso-octane were chosen as the endpoints for the octane scale because it was easy to get high-purity samples of these two hydrocarbons in 1927, when the scale was established.

An octane rating above 100 means that the fuel can withstand higher pressures than iso-octane before autoigniting. Ethanol, for example, has an anti-knock index of 110. A rule of thumb is that the octane rating increases as the number of carbon-carbon bonds in the molecule increases. Branched and ring molecules generally have higher octane ratings than straight-chain molecules.

Octane only tells you how much pressure the fuel can withstand before autoigniting, not how much energy it contains. A fuel's energy output depends on the difference between the energy contained in the chemical bonds prior to combustion and the energy contained in the re-formed chemical bonds after combustion. In general, the higher the ratio of hydrogen atoms to carbon atoms in a molecule, the more energy the molecule will release.

Five gallons of fuel contain more energy than two gallons of the same fuel because there are more fuel molecules in five gallons of fuel than in two gallons. We usually talk about fuel in terms of energy density, or how much energy is contained in one gallon of fuel. One gallon of gasoline contains energy equivalent to 31,000 food calories, which is about 115 McDonald's hamburgers. There is a rough correlation between octane rating and energy density within the hydrocarbon family; however, one gallon of ethanol has only 63 percent of the energy in one gallon of gasoline, even though ethanol has a much higher octane rating than gasoline. Surprisingly, the energy density of pump gasoline turns out to be pretty similar to the energy density of the Sunoco gasoline used by NASCAR. A NAS-CAR engine, however, combusts more fuel molecules at a time.

Not only do NASCAR engines make more energy per power stroke, they make power strokes more often than street engines because they run much faster than street engines. A typical NASCAR engine reaches 9,500 rpm at most tracks, while your engine happily putters along at 2,000 or 3,000 rpm.

Rotating a crankshaft at 9,500 rpm isn't too hard, but remember that the valves have to open and close just as quickly. Andy had a cylinder head and some valves in his office, which he used to demonstrate how the valve train—everything associated with opening and closing the valves—works. Each valve in a 2,000-rpm engine opens and closes a thousand times per minute, which is almost seventeen times each second. Each valve in a 9,500-rpm engine opens and closes seventy-nine times each second. In two hours, the racing valve will hit its valve seat 568,800 times. Your 2,000-rpm engine valve will hit just 120,000 times in the same period—and nowhere near as hard, Andy noted. The cam lobes on a production engine are shaped to slow the valves as they close so that the valves hit the valve seats more gently. NASCAR engines don't have time for that luxury, Andy said, clanging the valve against the seat a few times. The valves must stay open as long as possible, which means opening and closing them as quickly as possible.

High speeds require every part of the valve train to be extra strong. Andy brought out a steel pushrod from a Corvette, which is about seven and a half inches long and a little more than a quarter inch in diameter. He then handed me a pushrod from one of BDR's engines. The racing pushrod is tapered on the ends and the middle is thicker than my thumb. It is almost inconceivable to think something this sturdy could break, but a high-speed movie of a pushrod in action shows that it flexes like a ski pole.

There is one place where the stock Camry engine has an advantage: It has four valves per cylinder, which helps move gas through the engine faster. NASCAR allows just one intake and one exhaust valve per cylinder. A typical racing intake valve is a little more than two inches in diameter, and a typical exhaust valve about 1.6 inches in diameter. I would have described the valve as being shaped like a Hershey's Kiss, but Andy pointed out that the valve is much more complex, with very precisely defined angles on the top and bottom of the valve end.

Andy held up the exhaust valve to show me its large, rounded bottom. I guessed that the extra material was there to make a better seal, but when he inserted the valve into the head, it was clear that the bottom of the valve sticks out into the chamber and never touches the valve seat. Andy explained that the rounded shape helps exhaust gases flow out of the chamber more efficiently. A sharp edge would cause turbulence and slow the gas flow. This is the level of detail that teams must consider to be competitive.

Production car engines use steel valves because they are inexpensive and good at dissipating heat. Although NASCAR teams can use steel, most use titanium valves because they have less inertia than similarly sized steel valves. Titanium valves are strong and light, but also expensive. A steel valve might cost eight dollars, while a titanium valve runs about sixty bucks. Since the engine gets new valves for each race, that's about a thousand dollars per race just in valves.

Valves transfer heat to the valve seats when they close, which they do for about seven and a half thousandths of a second each cycle. Titanium valves don't transfer heat as well as steel, so special valve seats are required. The valve seats have to dissipate heat, but they also must stand up to repeated valve closings. Unfortunately, the things that make a material hard (like defects) reduce its ability to dissipate heat.

NASCAR engines have beryllium-copper valve seats, which dissipate heat better than steel or cast iron (or titanium). A mixture of copper and a few percent beryllium is cooled rapidly to disperse the beryllium atoms throughout the copper. The alloy is then heated to form small beryllium-copper grains that increase the material's hardness with minimal decrease in heat transfer ability. A harder material can be used on the intake valve seat, but the exhaust valve seat, where the exiting gases can reach 1,300°F, uses a softer alloy that has better heat dissipation properties.

The ring-shaped valve seats are inserted in the cylinder head by cooling the seat and heating the head. The valve seat shrinks a little, the space in the head for the valve seat gets a little larger, and the valve seat is held securely in place when the materials return to the same temperature. A similar trick can't be used for removing the valve seats, however, because both metals expand when heated. Andy pointed out one of the BDR machinists using a milling machine to remove the old valve seats from an aluminum cylinder head.

The valve and the seat become very hot during operation. The valve strikes the seat with enough force that tiny bits of the valve seat can be pulled away when the valve opens. The valve-seat atoms stick to the valve or are vaporized by the heat and discharged with the exhaust. Over time, if enough material is removed, the valve seat gets smaller (called valve-seat recession) and the valve can't close completely. Once exhaust gases leak around the valve during the high-pressure, high-temperature part of the cycle, the gasoline just burns, and as Andy put it, "everything in there just melts."

The main protection against valve-seat recession used to be leaded gasoline, which gets its name from added tetraethyl lead (TEL) molecules. TEL was discovered in 1921 and used as an anti-knock additive. Adding TEL to gasoline increased its octane, but also deposited a thin coating of lead that lubricated the valve seats and heads, which discouraged sticking and slowed valve recession. Concerns about lead contamination and lead's propensity for clogging catalytic converters prompted the U.S. government to ban lead from consumer gasoline in 1973. Although aviation and racing were exempted, an alternative fuel was desirable—but only if valve seat recession could be minimized.

Unleaded fuel made its first appearance at the NASCAR Sprint Cup Series level in February 2007 at Fontana. The new fuel (Sunoco 260 GTX) has a different mix of hydrocarbons than the old leaded fuel. I was surprised to find that the new fuel has an anti-knock index of only 98, because I'd always read that an engine with a 12:1 compression ratio needs at least 108 octane gasoline. Andy clarified this for me by noting that the octane required is not determined by the compression ratio, but by the actual pressure in the chamber prior to the spark. The valves stay open for so long on a NASCAR Sprint Cup engine that the intake valve hasn't closed entirely by the beginning of the power stroke. Consequently, the engine cylinders don't reach the pressures you'd expect for a true 12:1 compression ratio.

NASCAR never did find an additive to replace TEL in protecting against valve-seat recession. "We have to deal with it on the design end," Andy shrugged. One solution is coating the valves and/or seats with very thin layers of friction-reducing, thermally insulating, and/ or wear-resistant coatings. Andy showed me a valve with three different types of coatings along the valve end and stem. Hard coatings like chromium nitride, titanium nitride, and diamondlike carbon are used to decrease wear. Molybdenum-based coatings are used to help the valves slide smoothly. Some teams use ceramic coatings, which are thermal insulators, to protect the combustion side of the valve from high temperatures. Coatings have become so important, Andy said, that many valve suppliers have their own coating equipment on site.

The different mix of hydrocarbons in the new fuel requires a different mix of fuel and air. The hydrogen-to-carbon ratio changed from 2.1:1 in the old fuel to 1.75:1 in the new, which means that more fuel has to be let into the cylinder to achieve the right ratio of gasoline to oxygen molecules. Although production cars use fuel injection, NASCAR engines mix air and fuel using a carburetor. Fuel injection would allow better performance, but fuel injectors pose a challenge for NASCAR to monitor.

Carburetors work in a simple—and strictly mechanical—way. Air enters through the cowl—the area just beneath the windshield— and passes through an air filter before flowing into the carburetor through a narrow tube. The narrower the tube, the faster the air flows and the lower the air pressure. Fuel is sucked into this low-pressure region through very small holes and mixes with the air. Adjusting the sizes of the holes to vary the fuel/air ratio is called "jetting" the carburetor. If there is too much fuel for the amount of air, the engine runs "rich," and if there is too much air for the amount of fuel, the engine runs "lean." Neither is desirable.

The combustion equation is demanding. You must have *exactly* twenty-five oxygen molecules match up with *exactly* two octane molecules. If you have three octane molecules for the twenty-five oxygen molecules, the extra octane molecule is wasted because it has no oxygen partners.

NASCAR controls how much fuel can be used per cycle by limiting the amount of air coming into the engine. All cars use the same Holley carburetor, which lets a maximum of 830 cubic feet per minute of air into the engine. This restriction keeps speeds in a safe range—except at the two longest tracks on the NASCAR circuit.

Daytona and Talladega have high banking and very long straightaways, which means high speeds. Bill Elliott set a Talladega qualifying record of 213 mph (343 kph) in 1987. During the race that followed, Bobby Allison's rear tire blew and his car became airborne. Although Allison walked away from the accident, NASCAR realized that the speeds at Talladega and Daytona had become too high.

NASCAR introduced the restrictor plate the next week. A restrictor plate is a small, 1/8"-thick metal plate with four holes. The restrictor plate goes between the carburetor and the intake manifold, further limiting how much air gets into the combustion chamber. Restricted engines use a modified carburetor, different intake, smaller ports, and a smaller camshaft. Restricted engines have about 350 to 400 less horsepower, which reduces speeds by about 40 mph, depending on the size of the holes in the restrictor plate (diameters have ranged from 7/8" to 1" over the years). NASCAR makes the holes as large as safely possible, which is why they sometimes use odd sizes like 57/64". Decreasing the hole diameter by 1/16" takes about 50 horsepower away from the engine. The power is so sensitive to the hole size that NASCAR has an extensive quality-control program to ensure that all competitors get restrictor plates with the same size holes.

Restricted engine power means that the driver must hold the

throttle wide open almost all the way around the track. Most drivers find this less than exciting. Michael Waltrip suggested that driving at Daytona could be done by "an intoxicated orangutan." Another reason many drivers do not like restrictor-plate races is because the limited horsepower keeps the field tightly bunched. One little bobble can lead to the "Big One": a crash that potentially takes out half the field. Bobby Allison noted that the alternative, however, is that "if we left the engines unrestricted, we'd have cars going 240 mph and they'd be landing forty rows into the grandstands."

With all the limits on engine dimensions, making engines that run at higher speeds was one of the few ways engine designers could gain an advantage. Maximum engine speeds were around 7,000 rpm in 1970, but rose to 10,000 rpm by 2004. The problem with ever-increasing engine speeds is that the materials used in the engine—especially in the valve train—have to get lighter and stronger, which means they also get more expensive. A $60 titanium valve becomes a $250 valve when made from a lightweight, high-strength titanium alloy. That's $4,000 per race just in valves.

NASCAR found a clever way to put a lid on escalating engine speeds. They didn't make a rule limiting engine speeds (which would be hard to enforce), but instead implemented a gear rule. How fast the engine runs is directly related to how fast the car goes. If the engine runs at 9,500 rpm and the car is going 180 mph, the tires rotate 2,160 times per minute. There is no way to connect the engine directly to the wheels without slowing down the rotation rate. Decreasing the engine speed decreases engine power, so the rotation rate is slowed using two sets of gears—the transmission and the rear-end gear—that mechanically reduce the rotation speed transmitted to the wheels while still allowing the engine to work at its optimal speed.

Gears (wheels with teeth) are used in pairs, with one gear driving the other. If the first gear has sixteen teeth and the second has eight teeth, that combination has a two-to-one (2:1) gear ratio. The gear with eight teeth makes two revolutions for every one revolution that the larger sixteen-tooth gear makes. The larger gear rotates more slowly than the smaller gear because it makes fewer revolutions over the same time.

The input shaft of the transmission is driven by the crankshaft and the output shaft connects to the drivetrain. The transmission contains pairs of gears that couple the input and output shafts. The rotation-rate reduction is fixed by the gear ratios, so multiple pairs of gears are needed to allow the wheels to rotate at different speeds while the engine stays as close to its maximum power band as possible. The four-speed transmissions required by NASCAR have five sets of gears, the fifth gear being reverse.

Lower gears, which are used at lower speeds, have to compensate for larger rotational speed differences, so lower gears have higher gear ratios. In a 2007 five-speed manual transmission Toyota Camry, the gear ratios are 3.54 (first), 2.05 (second), 1.33 (third), 0.97 (fourth) and 0.73 (fifth/overdrive). In first gear, the engine makes about three and a half revolutions for every revolution of the transmission's output shaft. In overdrive, the driveshaft rotates faster than the engine. NASCAR transmissions are not allowed to have gear ratios of less than 1.

The transmission decreases the rotational speed, but not enough to match the required rotational speed of the wheels. The rear-end gear is a ring-and-pinion gear that, in addition to changing the rotation rate, changes the direction of the rotation so that it is parallel to the rear axle. The gear rule NASCAR implemented in 2005 gives teams two (or sometimes three) rear-end gear choices at each track. Since the gear ratios determine the optimal engine speed, limiting the rear-end gear effectively limits the engine speed.

A "650 gear," which is typical for a track like Martinsville, has a gear ratio of 6.50:1. One rotation of the axle requires six and a half rotations of the driveshaft. Gear ratios range from 2.73 to 6.50, with lower gear ratios used at higher-speed tracks. Andy could design an engine to run at higher speeds, but it wouldn't improve the car's lap times because the required gears would force the engine to operate in an unfavorable rpm range.

When stock-car racing first started, the focus was on engines. When aerodynamics was recognized as important, many teams switched their efforts to understanding how the car's shape affected its speed. Most of the easy gains in the engine had already been made, or so it seemed. The fixed body shape of the new car just might bring attention back to the engine, which clearly makes the people working in the engine departments happy.

"After all," Andy laughed, "the car's just a dolly to move the engine around."

Six

The Wizard and the Flying Car Problem

In a sport where people have nicknames like "Oil Can" and "Fatback," being known as "The Wind Wizard" is quite a compliment. Gary Eaker does look something like a wizard, with his mustache and shaved head. His shorts and casual shirt make him a little less intimidating as he sits behind his desk at the AeroDyn wind tunnel in a business park just opposite the Carolina Brewery in Mooresville.

While a real wizard may view getting a car to fly as an accomplishment, one of Gary's most significant contributions to NAS-CAR is helping keep race cars firmly on the ground. Gary may not be an actual wizard, but aerodynamics—the study of how air moves—can seem like wizardry, because air is both complicated and (usually) invisible.

The motion of air is complicated because the atoms or molecules in fluids—and air *is* a fluid—are only loosely connected to each other. Describe the path of a thrown baseball. The atoms in the baseball maintain the same positions relative to one another, so the ball moves through the air as a single unit. Now describe what happens if you throw a handful of slime. The motion of slime is more difficult to describe because slime doesn't have a fixed shape. Slime spreads out and even comes apart when you throw it.

At least you can see slime. People say that Dale Earnhardt, Sr. could "see" the air because he was so expert at using aerodynamics on the track. Most of us don't have that ability. The irony, Gary said, is that "I avoided electrical engineering and went to mechanical engineering because connecting rods and crankshafts are physical things I can comprehend, not like electronics." He threw up his hands in mock frustration. "Then I end up working with air."

Those of us who didn't study aerodynamics at the "University of Daytona" have two tools to help us "see" how air interacts with cars: wind tunnels and computers. Computers are relatively recent, but wind tunnels have been used to study aerodynamics since 1871. The Wright brothers used a scale-model wind tunnel in 1901 to help design their first planes.

Wind tunnels have advanced quite a bit since then. Wind speeds range from just a few miles per hour to more than 20,000 mph in wind tunnels that are used to study airflow around ballistic missiles and spacecraft. The largest wind tunnel in the world belongs to NASA and can test planes with wingspans of up to one hundred feet.

Gary's experience with wind tunnels started at General Motors. After earning a bachelors degree in mechanical engineering from General Motors Institute (now Kettering University) in 1976, he worked as a project engineer with GM and moved into race-car aerodynamics in 1981. He worked at Hendrick Motorsports from 1994 until the end of 2001, when he left to build the AeroDyn wind tunnel.

At first glance, a wind tunnel is just a big room with a very powerful fan. A closer look reveals much more. A bank of twenty-two individually speed-controlled, 100-horsepower electric fans stands about eighteen feet high and consumes 1.6 megawatts of electrical power—equivalent to 16,000 100-watt lightbulbs. The AeroDyn wind tunnel reaches a maximum wind speed of 130 mph, which is equivalent to a strong Category 3 hurricane.

Creating hurricane-force winds indoors isn't cheap. An hour at AeroDyn costs about $1,500, but wind tunnels are an important enough tool that AeroDyn runs three shifts Monday through Friday, and they are almost always booked.

"The race teams basically can't get enough of the wind tunnels," Gary explained. He recently opened a smaller wind tunnel, called A2, which he refers to as "an oversized science fair project." A2, he hopes, will make wind tunnels accessible to race teams without NASCAR-size budgets.

A wind tunnel can let you see—literally—how air moves over your car. The array of fans produces a smooth flow that acts like sheets of air stacked on top of each other. A wand releases smoke in front of the car and the smoke particles show the paths—called streamlines—that the air molecules follow as they are deflected around the car. Tufts of string can also be attached to the car to show the airflow.

These are visual ways of understanding airflow. Real understanding requires numbers. Pressure gauges measure how hard the air hits each area of the car. The car sits on a scale in the wind tunnel that is sensitive enough to detect the change in weight when a half-dollar is placed on the car's front bumper.

"It's a slight exaggeration to say that we can use it as a letter scale," Gary smiled, "but not much of an exaggeration."

A car's weight doesn't actually change in the wind tunnel; however, every time an air molecule hits the car, it exerts a force. Some of those air molecules push downward, making the car appear heavier. The force that air exerts on the car's body is proportional to the air pressure times the area the air acts over. A larger area produces more force.

We can understand why the pressure varies over the car by tracking the air molecules' motion. The air molecules start to bunch up when they first approach the car's front bumper, like traffic on an expressway slows down when people have to change lanes because of a stopped car. Once the air molecules get past the car's nose, they speed up as they travel over the hood, but then they hit the windshield and slow down again. After traveling up and over the windshield, the air molecules gain speed and flow freely along the roof.

While the front of the car pushes air away, the back of the car allows the air to rush in to fill the hole made by the front of the car. If you stir syrup with the back of a spoon, the spoon first breaks though the syrup, and as soon as the spoon passes, the syrup rushes into the space left by the spoon. A speeding car has the same effect: It makes a hole in the air faster than the air molecules can fill it. The result is turbulence—the same turbulence you experience on airplanes. Instead of moving in nice smooth sheets of air, turbulent air molecules move in chaotic, vortexlike paths, as shown below. You can see the transition from smooth to turbulent in smoke rising from a smoldering fire or a cigarette. The smoke starts rising in straight lines, but becomes turbulent the farther away it travels.

Air flow →   Turbulence

On the next page, I've drawn the airflow over a generic race car, with longer arrows indicating faster-moving air (top illustration). The air is pushed away from the car in the front, starts becoming turbulent near the rear window, and hits the wing before becoming turbulent again behind the car. Faster-moving air creates less pressure.

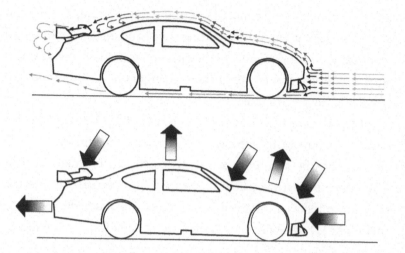

Although I've drawn the picture in two dimensions, the air exerts pressure in three dimensions. Forces up and down correspond to lift and downforce, while forces along the length of the car are called drag forces. Forces acting perpendicular to the length of the car are called side forces, and these would move into or out of the paper in this illustration. Most forces push in more than one direction, as the bottom illustration shows, because the air exerts pressure perpendicular to the surface over which it passes. The force on the windshield, for example, pushes down and toward the rear of the car.

Drag forces, which act along the length of the car, always oppose the car's motion. The air molecules hitting the front bumper push backward, and the lower-pressure region created by turbulence behind the car pulls the car backward. The size of the drag force depends on how fast the car is moving, the air density, and the car's cross-sectional area and shape. A faster car experiences more drag because it has to push air molecules out of the way faster. Denser air increases drag because there are more air molecules hitting each area on the car. A larger cross-sectional area increases drag because more air molecules have to be moved out of the way. You can test this last effect yourself by (carefully!) holding your hand out a car window. If you hold your hand parallel to the ground, its cross-sectional area is smaller and you feel less drag than when you hold your hand perpendicular to the ground.

The final element contributing to drag is the car's shape, which determines how easily the car "cuts" through the air. The shape is characterized by a number called the coefficient of drag, which is determined primarily by how air detaches from the car. The easier it is for air to follow a car all the way around, the lower the car's drag coefficient. Drag coefficients range from 0.04 for a streamlined, winglike shape to 1.15 for a short cylinder with the circular end facing the wind. The two shapes in the picture below deflect the air similarly in the front, but the back of the sphere curves too quickly for the air to follow. Air remains attached to the lower object longer, which creates less turbulence behind the object.

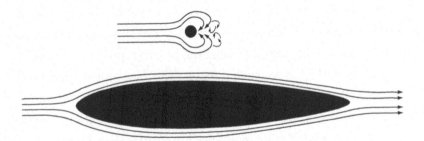

Drag is due to friction between the molecules and the car, so rough surfaces like grilles are taped over when possible, and nothing sticks out from the car. Drag is also produced by little tornados called vortices. Vortices are produced at the edges of airplane wings or at the rear of the car when air leaves the spoiler in such a way that it rotates.

Sharp edges encourage air to separate from the car, while curved surfaces help the air stay close. Race cars have sharp corners where the sides of the car meet the rear bumper cover. This prevents air coming down the side of the car from being pulled around the back and creating drag.

The windows and wheel openings are good examples of how car shape minimizes drag. The curvatures of the A- and B-posts around the windows minimize the air entering the cockpit and help pull out any air that does get in. The window net on the driver's-side window is pulled taut because even though the net has holes in it, a taut net provides a smoother surface for the air to flow over than it would if it were floppy. The fronts of the wheel openings flare outward and have sharp edges to encourage the air to travel out and around the wheel, instead of into the wheel opening. The back sides of the wheel openings are bent slightly inward to help any air that does get into the wheel opening to escape.

Drag coefficients of non-speedway NASCAR cars are around 0.35, which is larger than the drag coefficients of passenger cars like the Toyota Camry (0.28) or Ford Fusion (0.33). Indy cars typically have drag coefficients around 1.00, because of their wings and because they don't have fenders to reduce drag around the wheels.

The effect of drag on race cars was fairly well-known, but Gary explained that the demand for wind tunnels really picked up when stock-car designers started appreciating the importance of the forces that act up and down on the car. People who designed airplanes already appreciated these forces. When air molecules encounter a wing, some air molecules go over the wing and some go under. If the force under the wing is larger than the force over the wing, there is a net force pushing up, which is called lift. If the force pushing down is larger than the force pushing up, the net force is down. Airplane designers call this negative lift, but race engineers call it downforce.

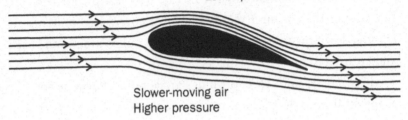

Faster-moving air
Lower pressure

Slower-moving air
Higher pressure

Slower-moving air exerts more pressure on the wing than faster-moving air. I was taught that lift is produced only if the top of the wing is more curved than the bottom. The argument goes like this: Assuming that air molecules have the same speed just before they encounter the wing and just as they leave the wing, the air molecules that go over the top have to travel faster than the air molecules that go beneath if the air molecules are going to meet up on the far edge of the wing. The force on top of the wing is thus less than the force on the bottom of the wing, which creates lift.

Although this argument can be found in many textbooks, it is not correct, because the assumption on which the explanation is based is unfounded: There is no reason why two air molecules that were next to each other before encountering the wing should meet up again after they go around the wing. What we *do* know is that the molecules that go over the top of the wing do travel faster than the molecules that go under the wing, and this is what produces lift; however, if you assume (incorrectly) that the air molecules need to meet up again after the wing, the lift you calculate is much smaller than the value you actually measure.

Lift and downforce are created primarily on horizontal or near-horizontal surfaces. Air molecules move quickly over large-area surfaces like the hood and roof of a car, creating low-pressure regions.

Any place that forces the air to slow down, like the windshield, creates higher-pressure regions.

Race-car designers figured out in the 1960s that wings could be used not just to create lift, but also to create downforce, and downforce gives the car more grip by pushing the tires harder into the track. Open-wheel designers put wing-shaped appendages on their cars to create downforce, but stock-car designers realized that the car's entire body could be used to make downforce.

Where downforce is created is just as important as how much downforce is created. The scale in the wind tunnel measures more than just the car's total weight: There are six separate weighing points, which allow the aerodynamicist to determine exactly *where* the forces act. If the front of the car "weighs" more than the back, for example, there is more front downforce than rear downforce.

The up and down forces are characterized by a lift coefficient the same way the drag coefficient characterizes the forces opposing the car's forward motion. Passenger cars have lift coefficients from +0.10 to +0.40, while a stock car may have a lift coefficient around -0.42. Indy cars have lift coefficients that range from -2.00 to -3.00. A negative lift coefficient means the force is pushing down. Although passenger cars have positive lift coefficients, they are small. Besides, you shouldn't be driving a passenger car so fast that you generate enough lift to offset the car's weight.

Stock cars have negative lift coefficients, but there is plenty of evidence that they can—under certain conditions—leave the ground. Although Gary's "Wind Wizard" nickname is due in part to his skill using aerodynamics to make fast race cars even faster, some of his most important aerodynamic innovations are found on every NASCAR Sprint Cup Series car racing today, regardless of its manufacturer or owner.

Although most of the air travels over the car, some air can get beneath the car. One of the reasons that stock cars have what is known as "rake"—that is, the back end of the car is higher than the front end—is to help air escape from under the car. If the pressure under the car is greater than the pressure on top of the car, the car will have lift. The splitter on the new car rides as close to the track as possible, which minimizes how much air gets under the car; however, there are other ways a car can generate lift.

The yaw angle is the angle between the direction the car is heading and the direction of the air flow. A car pointing in the same direction as the air flow has zero yaw. During normal racing conditions, a car usually has a few degrees of yaw; however, if the car spins, its yaw angle can reach 90 degrees or more. A 90-degree yaw angle means the car is traveling sideways, and if you look at a race car from the side, its shape isn't all that different from a wing. Air moving quickly along the roof, hood, and deck lid can generate enough lift to make the vehicle airborne.

Two of Gary's first aerodynamic safety contributions were the recessed right-side window and the right-side rocker skirt. Cars turn left on oval tracks and are more likely to spin with their right sides facing in the direction they are moving. The recessed right-side window, which was introduced in 1988, creates a sharp edge on the top of the window that forces air to separate from the car instead of flowing quickly over its roof. Gary also designed a rocker-panel skirt that runs along the bottom of the car's right side to prevent air from getting underneath during a spin. These improvements helped decrease lift, but they didn't solve the problem entirely.

"Somewhere along the way," Gary continued, "Ford got on board. They were the ones who came up with the roof strips." Roof strips are thin pieces of metal about ½" high that run front-to-rear along the edges of the roof. Like the recessed window, roof strips disrupt airflow and decrease lift. Again, the roof strips helped, but didn't solve the problem completely.

"Then we saw that when you got (a yaw angle) beyond 110 degrees, some of the lift came back as the air reattached," Gary said, using his hands to illustrate. "As you're coming at an angle, you're generating a vortex, and the vortex may help hold the air onto the roof." Essentially, the air breaks over the top of the car and creates a vortex parallel to the airflow. The rush of air over the roof creates lift.

In the late eighties and early nineties, Gary started thinking about ways to decrease roof lift. He found that a metal plate about six inches high standing up on the roof successfully separated the air from the car even at high yaw angles. Although it worked, it wasn't an acceptable solution.

Gary explained that safety devices have to satisfy a number of requirements. A safety device must work reliably, even under extreme conditions. Teams shouldn't be able to exploit the device to gain speed. Conversely, if the safety device slows down the car, teams are likely to try to find a way around it. Gary's vertical plate fell in the last category.

"That's as far as we took it at the time because it was obvious that you weren't going to go dragging this plate around on the roof," Gary said. Another development—the porous roof—had to be abandoned for the same reason. An array of tiny openings in the top layer of the roof increased drag by creating additional friction with the air. That was discarded, Gary explained, because the porosity slowed down the car. "The question became, what are these guys going to do to fill in the pores? And now you're back to square one."

They needed something that would only slow down the car when the car needed to slow down. During a brainstorming meeting at GM in 1993, Don Taylor came up with the idea of having the deck lid (trunk) deploy upward and rest on the back window. The deck lid would have the same effect as Gary's vertical plate, but would only deploy if the car's yaw angle became large enough. The first test was conducted at the Darlington airport and used the NASCAR corporate jet to generate 200-mph winds. The idea worked when the deck lid deployed in just the right position, but the deck lid didn't always deploy in the right position.

"So it was right there and right then," Gary said, "I was talking with Gary Nelson (then NASCAR's managing director of competition) and it was the proverbial lunchtime on a napkin." (A hallmark of most scientific-discovery stories is that the solution is developed on either a napkin or the back of an envelope.) Gary realized that the problem was ensuring that the deck lid always deployed at the correct angle.

Gary continued. "I said, 'We have to define these edges relative to the roof and, just like the flaps on an airplane, have something that deploys right there.' Because of the previous work with the plates, I said, 'I think they're going to have to be this tall and in this location.' At that point Jack Roush said, 'We can manufacture these things.'" Roush made a prototype from Gary's design and they went into the wind tunnel to perfect the geometry. They knew the problem area was the roof, but they didn't know exactly where the lowest pressure area on the roof was.

"I'd been working in wind tunnels since '77," Gary said, "but I underestimated the complexity of the problem because we didn't go back to the most fundamental thing, which was doing a basic experiment on the pressure profile of the roof of the car. So we struggled and we struggled and we had a lot of false starts, but when we finally put pinholes and pressure taps over the entire upper surface, everything clicked."

They measured the pressure every ten inches horizontally and vertically on the front of the deck lid, the entire rear window and roof, and the rear part of the windshield. The results made it clear where to locate the deployable flaps. They also found a low-pressure region along the left-hand side of the rear window, which is why today's cars have a "shark fin"—a thin metal strip sticking up perpendicularly along the left-hand side of the rear window—that disrupts the airflow in that potentially troublesome area.

The eventual solution was adding two carbon-fiber roof flaps, each twenty inches wide and eight inches tall sitting in hinged carbon-fiber trays recessed into the car's roof. The flaps are located near the back of the roof, with the left flap perpendicular to the car's length and the right flap at a 45-degree angle to the first flap. The right flap is angled because cars tend to spin their rears toward the right and the right-most flap will deploy as soon as the yaw angle exceeds the critical value.

The air under the flap is normally at the same pressure as the air over the flap. If the air pressure over the roof starts to decrease, the air pressure under the flap becomes greater than the air pressure over the flap and the flap starts to rise. The wind catches the now-exposed edges of the roof flaps and pulls them upright (just like the original metal plate), which creates turbulence that decreases lift. Two cables control how far the flaps rise—you can sometimes see the cable ends hanging down in the in-car shots. Keeping the car on the ground during a spin scrubs off speed by maintaining friction between the tires and the track.

This story illustrates how research really works. The search for a way to minimize roof lift started in 1987. Different teams of people developed ways to decrease lift, but none completely solved the problem. Some ideas (like the porous roof and the metal plate) had to be abandoned because even though they worked, they were impractical; however, the information learned from these attempts turned out to be useful in developing the ultimate solution of the roof flaps. The final design validation was completed in January 1994, and NASCAR required roof flaps on every car for the Daytona 500 the next month.

"At the end of the day," Gary said, "it was a long, evolutionary process that started with the side windows and ended with the roof flaps."

Although roof flaps have dramatically decreased takeoff accidents, the problem is not totally solved. Despite having roof flaps, Ryan Newman at Daytona in 2003, Elliott Sadler at Talladega in 2003 and 2004, Tony Stewart at Daytona in 2006, and Kyle Busch at Talladega in 2007 all managed to get their cars into the air. NASCAR analyzes each incident to determine if safety equipment malfunctioned or didn't behave as expected, and whether other factors were involved. Many of these accidents had extenuating circumstances, such as getting a lift upward from another car, running up against the wall, or getting one of the car's wheels caught on the interface between the grass and the track.

Drivers will doubtlessly discover other ways to make stock cars act like airplanes, but anticipating new problems is what makes science interesting.

"The root word for engineer doesn't come from engine," Gary smiled. "It comes from ingenuity."

Seven

Running with the Pack

Junior Johnson didn't have a ride for the 1960 Daytona 500, despite having won five races the previous year. A little more than a week before the race, Johnson got a call from car builder Ray Fox, who had received last-minute sponsorship to put a car in the race and desperately needed a driver. Johnson wasn't exactly enthused: Pontiac had the fastest cars and Fox was not only running a Chevrolet, he was running last year's Chevrolet. Despite their best efforts, Johnson qualified 22 mph slower than pole-sitter Cotton Owens.

Johnson was surprised during practice to find that he could keep up with the Pontiacs—but only if he stayed very close on their rear bumpers. Johnson went out on the track alone, and then trailing the Pontiacs. Alone he was slow, but when he got right up on the rear bumpers of the Pontiacs, "They couldn't shake me," Johnson said. "I knew then I was right about the air creating a situation—a slipstream type of thing—in which a slower car could keep up with a much faster one."

Johnson started ninth and planned his pit stops around those of the Pontiacs. Bobby Johns passed him for the lead in lap 170 of 200. Johnson got as close as he could to Johns and then got dramatic confirmation of the power of air.

"Then, coming off the second turn with ten laps to go, one of the damndest things happened I ever saw on a track," Johnson recalled. "The back glass popped out of Bobby's car and flew into the air. I think our speed and the traffic circumstances combined to create a vacuum that sucked that back glass right out."

Johns spun and Johnson won. Other drivers quickly caught on to what Johnson was doing and tried it themselves. The phenomenon was christened "The Draft." Junior Johnson didn't "discover" aerodynamics—watch a flock of birds and you'll see that they intuitively know how to draft—but he was the first to figure out how to use it to get a competitive advantage on a race track.

Johnson's discovery came at an opportune time: Teams had run out of easy engine improvements. Once they started building cars from scratch instead of modifying production cars, large-scale changes in body structure were possible. The era of engines gave way to the age of aerodynamics.

Early aerodynamic experiments were guided first by wind-tunnel data and later by a new technique: computational fluid dynamics (CFD). Fluids like air are described by the Navier-Stokes equations, which are significantly more complex than the relatively simple equations that describe the motion of a ball. Solving the Navier-Stokes equations for real situations had to wait for the invention of powerful computers. Now, CFD is used to study problems like the flow of gases through engines, ocean-current circulation, the design of nuclear weapons, and the motion of planes, trains, submarines, boats, rockets, and race cars.

A CFD simulation starts by building a car in your computer. The car's surface is divided into small volumes called cells. Anything around the car—the ground or other cars—also has to be included in your computer model. The equations that describe airflow in each cell are solved numerically. The program calculates the speed and direction of the air molecules, which tells you how much and the direction of the force that air exerts on each surface. Simulations using smaller cells produce more accurate values, but take longer to complete because there are more calculations to do.

Anyone can buy a commercial CFD program and generate data; however, using that data to make your race car go faster requires the expertise of someone who understands CFD *and* race cars. Dr. Eric Warren, who served as the vice president and technical director for Gillett Evernham Motorsports (GEM), is such a person.

GEM, which is located in Statesville, North Carolina, was founded in 1999 when Ray Evernham left a successful career as Jeff Gordon's crew chief to spearhead Dodge's reentry into the NASCAR Sprint Cup Series. Eric started at GEM in January 2002 and, until his departure in 2007, was responsible for performance and manufacturing of all GEM race cars.

The engineering departments at race shops look pretty much like any other small engineering company: Carrels of identical desks and hutches are arranged in small groups on a durable, dark-colored, carpeted floor. Several young engineers, all wearing red company polo shirts, gathered around a large computer monitor displaying a wireframe model of a Dodge Charger and discussed the brightly colored splotches that indicated the pressure on different parts of the car's nose.

Eric, who oversees a crew of about thirty from a glassed-in office at the back of the room, apologized for his "messy" desk, which holds a laptop, four neatly stacked piles of paper, and a miniature Stanley tool chest holding paper clips and pens. His diploma and pictures of his wife and two children line the top shelf of his bookcase. The rest of the shelves hold textbooks ranging from Halliday and Resnick's *Fundamentals of Physics* to the Millikens' classic text on race-car vehicle dynamics. Eric, who sported a North Carolina State University sweatshirt, received his Ph.D. in aerospace engineering from that institution in 1997. He never expected he would end up working on race cars instead of rockets.

"My story's definitely strange," Eric said, the declarative nature of his sentences contrasting with his soft North Carolina accent. "Never had seen a race, never had an interest in races. I grew up an hour north of here, so I knew what NASCAR racing was, but I was not a fan." He was close to finishing his dissertation when a chance conversation with a childhood friend at a party resulted in a visit to the Kranefuss-Haas race shop. In a chance meeting "walking down the steps getting ready to go to lunch—I mean, at the front door walking out," Michael Kranefuss told Eric to call when he was about to graduate.

"I thought it was just a cordial thing," Eric said. "When I was a few weeks out from finishing my dissertation, I figured I'd just give him a call. So I came down and sat in his office and, like, two seconds into it, he asked, 'When do you want to start?'."

Eric realized Kranefuss' job would pay better than the government positions he had been considering, plus it would allow him to stay closer to home. He accepted the job. Roger Penske bought out Carl Haas's half of the business shortly afterward, and Eric ended up as the chief engineer of the combined teams.

"So within three months of getting into the sport, I basically had a lead engineer role with Rusty Wallace and Penske."

One reason Eric's skills are in demand is that CFD lets you try out almost any imaginable configuration without having to actually build anything. You can do things you can't do in a wind tunnel, like look at air flow around multiple cars. You're not limited by your wind-tunnel budget, but you do have to invest in pretty heavy-duty computing power. Simulating an entire car can take several days, and your results are only as good as the data you input and the programs you use. Some teams use commercial CFD programs, but most create their own programming to deal with the unique aspects of racing.

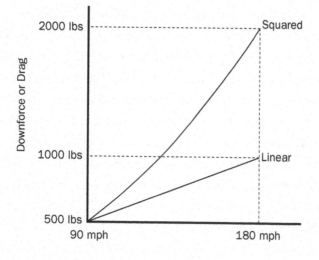

Aerodynamic considerations are different at almost every track, in large part because the size of the forces acting on the car depends on how fast the car is going. Drag and downforce increase with the *square* of the speed. I've compared linear and squared dependences on speed in the graph above.

Let's say you have 500 pounds of downforce when you're going 90 mph. If downforce depended linearly on speed, you would have 1,000 pounds of downforce at 180 mph; however, downforce depends on the *square* of the speed, so there are actually *2,000* pounds of downforce at 180 mph—four times the original 500 pounds.

This sensitive dependence on speed means that aerodynamics plays different roles at different tracks. When speeds are below about 120 mph, as they are at short tracks and road courses, aerodynamics is less important than heat management and mechanical grip. Cars that look ready for the scrap heap can—and have—won at tracks like Martinsville.

Intermediate tracks, where speeds are typically much higher, are sometimes called "downforce tracks," because cars really need the additional grip that downforce provides. At a track like Kansas, the old car could generate 1,400 to 2,000 pounds of downforce, which is significant compared to the 3,600 pounds the car and driver weigh.

Force is the product of pressure and area. The fender size and position determine how much force is created and in what direction. The old cars designed to run on intermediate tracks, as I mentioned earlier, looked like kidney beans, with the front and rear pulled toward the right side of the car. The left-front fender is much wider than the right-front fender because—as we'll see in the next chapter—it's harder to keep the left front tire on the track coming out of a turn. Tailoring each piece of the car body allows the car designer to shift where the forces act.

In the old car, the spoiler created most of the rear downforce. Shifting the rear of the car to the right kept the greenhouse from blocking air to the spoiler. Keeping the rear of the car as high as possible kept the spoiler in the airflow. Without enough rear downforce, the back wheels slide out from under the car, sending the driver spinning.

Drivers quickly learned how to take advantage of this balance. Ricky Rudd said, "If you want to pass a guy, you don't necessarily have to outrace him; you just drive up underneath him so close, and he'll *want* you to pass him."

Maneuvering behind another car allows you to change the way the air hits that car. If you get in just the right position, you can decrease the amount of air hitting the lead car's spoiler and cause more of the dirty (i.e., turbulent) air to tuck up under the car's rear end. "Taking air off the spoiler" decreases the rear downforce, and the driver on the receiving end is suddenly more occupied with keeping his car from spinning out than he is with staying in front of you. You just have to get close enough to let the air do the dirty work.

This technique has a potentially hazardous side effect. When you change your competitor's airflow, you also change how the air flows around your own car. You'll experience "aero push," which means that the car is "pushing" or not wanting to turn because you've lost some of your own front downforce. Elliott Sadler described it as feeling "like somebody has two jacks under the car and is lifting it off the ground. No matter which way you turn, it doesn't matter. It's just going to slide." Traveling in the wake of a car decreases front downforce because the wake draws air away from the trailing car.

Aerodynamics is equally important at superspeedways, but the emphasis shifts from increasing downforce to decreasing drag. The restrictor plates used at superspeedways reduce engine power by 350 to 400 horsepower. The drag force goes like the square of the speed, but the power needed to overcome drag goes like the cube of the speed. That means that if you double your speed, you get four times as much drag and you need *eight* times more power to overcome that drag. A car going 50 mph may only need 15 horsepower to overcome drag, but if the car doubles its speed to 100 mph, the same car needs 120 horsepower.

Minimizing drag is the number one priority at superspeedways, even if it means reducing downforce. Decreasing the drag coefficient of a car going 200 mph by seven thousandths effectively adds one full mile per hour to its speed. The old superspeedway cars were the thoroughbreds of race cars. The big, broad fenders of the intermediate car gave way to sleek streamlining with minimal grille openings and a gradually sloped front windshield that helped air flow over the car.

In contrast to intermediate-track cars, the goal for a superspeedway car is to keep air *off* the spoiler to decrease drag. Jimmie Johnson won the Daytona 500 in 2006, but he did it without his crew chief, Chad Knaus. The car had been set up so that changing the car height made the rear window bulge outward. The additional curvature decreased the amount of air hitting the rear spoiler and thus decreased the drag. (The infraction was discovered after qualifying. The car was modified and did pass inspection prior to and after the race.)

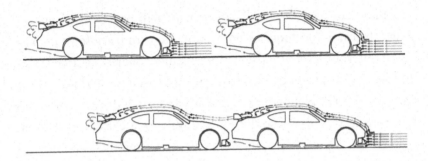

Superspeedways are where drafting is most important because of the restricted engines. Two cars drafting together can go 3 to 5 mph faster than a single car by itself. The duo can gain a little more speed by adding a third car, but they will gain only fractions of a mile per hour with each additional car.

The airflow creates a high-pressure region at the front of a car and a low-pressure region at the rear, as I've shown in the top illustration above. Both forces slow down the car. If you can put a low-pressure region immediately in front of your car, the car should go faster. This is what Junior Johnson discovered: The car in front of you has a low-pressure region behind it. Why not let that car part the air *for* you while you're pulled along in its wake?

When two cars get very close to each other, as I've shown in the bottom illustration, the air flows over both of them almost as though they were a single, very long car. Drafting allows the air to travel smoothly from the rear of the first car to the front of the second car without generating a lot of turbulence. I would have loved to have been there the first time Junior Johnson saw a CFD drafting simulation. I can only imagine how satisfying it would be to see your intuition confirmed by a supercomputer.

Drafting necessitates additional strategy. Let's say you're behind the leader in a line of cars at Talladega and you decide to pull out of line to pass. If one or more of the cars behind you pull out with you, you've stolen the draft from the lead car, and he will slow down. You and your partners, who still have the advantage of drafting, can pass. If the cars behind you think they are better off staying with the leader, no one follows you and you're "hung out to dry"—left without any drafting help and probably losing eight to twelve positions very quickly.

Stock-car drivers are natural experimenters. Following the example set by Junior Johnson, some drivers began to investigate what happens if, instead of just getting very, very close to the car in front of you, you bump into it. Because the first car is deflecting air, the drag on the second car is less, so the second car has enough engine power to go faster than the first car. When the trailing car bumps the leader, momentum is transferred to the first car. The first car moves faster and pulls the second car along.

"Bump drafting," as this technique is called, works very effectively, but only when executed properly. A car goes in the direction you bump it, which is fine on straightaways, but not so good in the turns, or when the bump comes at an angle. Improperly executed bump drafting may cause the front car to lose control, which is why drivers choose their drafting partners carefully.

The old superspeedway cars often had reinforced front bumpers, including extra support rods and even inch-thick steel plates. After the 2005 Daytona 500, NASCAR tried to put an end to what had morphed from "bump drafting" into "slam drafting" and mandated "soft" bumpers, which meant removing the metal plate and decreasing the front-bumper tubing diameter. They hoped that drivers wouldn't bump so hard if it meant they might damage the aerodynamics of their own cars.

Aerodynamics can also help drivers pass on superspeedways, which is notoriously difficult because of the reduced engine power. In a slingshot pass, the passing car comes from behind, on the outside of the first car, and then crosses through the wake behind the first car. The wake pulls the second car forward, allowing the driver to zip to the inside and pass the first car.

A car's aerodynamics is extremely sensitive to its shape, which (bolstered by access to wind-tunnel measurements and CFD simulations) led to teams developing a specialized car for every race. In addition to keeping the shop crews busy cutting off and replacing bodies, the emphasis on aerodynamics meant that an accident—even a relatively minor one—could change the car's aerodynamics enough to put you out of contention. Drivers and fans complained that the aerosensitivity made it harder for cars to pass each other on the track.

NASCAR first tried to limit the aerodynamics arms race by rules changes, but eventually, instead of trying to stuff the aerodynamics genie back into its bottle, they finally decided to just build a new bottle. The new race car shifts all the aerodynamic adjustability from the body to a rear wing and a front splitter. Instead of making a different body for each race, teams now build exactly one tightly legislated body style and adjust the aerodynamics for each track using the wing and the splitter.

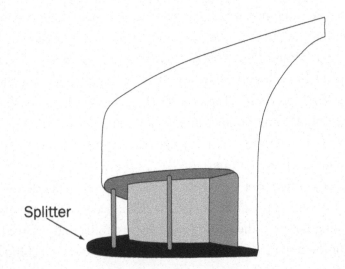

Splitter

The splitter shown in the drawing on the previous page is a half-inch-thick shelf at the bottom of the car's nose. The splitter is made from Tegris, a composite material that has about 70 percent of the strength of carbon fiber, but at about 10 percent of the cost.

The splitter is so named because it splits the air hitting the car's front: Some air goes over the car and some under the car. The splitter can be adjusted to have between zero and two inches protruding beyond the vertical part of the bumper. Exposing more of the splitter produces more downforce. Unlike the front valence it replaced, the splitter has a maximum height restriction: The top of the splitter can't be more than four and a half inches off the ground when the car is standing still.

Moving front downforce from the nose to the splitter allowed NASCAR to raise the front bumper and lower the rear bumper to make them the same height. A driver used to be able to stick the nose of the old car literally right underneath the car in front of him, which really disrupted the air flow. The matched-height bumpers on the new car prevent the use of this strategy and protect the fuel cell better. The nose on the new car is much more vertical, and cooling air comes in underneath the bumper (just above the splitter) instead of through the front grille, which hopefully will reduce debris clogging the air intake and causing overheating.

The nose is shorter on the new car and the cockpit is two and a half inches taller, which makes the windshield more upright. The more steeply sloped windshield increases the pressure in the cowl area, where the engine's air intake is located. The new car has a larger cross-sectional area than the old one. You'd expect that change to increase drag; however, NASCAR found that the larger greenhouse blocked air from getting to the spoiler, which actually decreased drag—and downforce. Raising the deck lid the same amount as the roof created additional drag that would require about

150 horsepower to overcome if the old spoiler were used. Switching
to a wing made the drag closer to that of the old car.

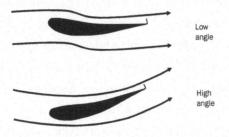

Low
angle

High
angle

The wing, which is shown above in cross-section, is about a foot long
and fifty-four inches wide, and made from carbon-fiber composite.
The wing can be rotated on its mounting brackets from an angle of
zero to sixteen degrees to vary the amount of drag and downforce. A
flatter wing produces less downforce (and less drag), while a larger
angle produces more downforce. A thin, L-shaped piece of metal
about ³⁄₁₆" (5 mm) wide sits at the trailing edge of the wing. This
Gurney Flap (also called a wickerbill) increases downforce with a
minimal increase in drag. A one-inch wickerbill is used at superspeedways.

   Although NASCAR issues wings to teams at the track, the
teams provide their own end plates, which are essentially wings
turned vertically to control side force. Teams can choose between
flat and cambered end plates, with flat end plates producing less
drag and cambered end plates producing more side force. Teams
may choose to use wickerbills on the end plates as well.

   On average, the drag on the new car is about 10 percent greater
and the overall downforce about 10 to 15 percent less than on the
old car; however, *where* those forces act is important. General Motors' CFD simulations show that the high-pressure areas on the old

car were spread out along the front and top edge of the nose, while the pressure on the new car is concentrated almost entirely near the splitter. Similarly, the high-pressure region spreads across the deck lid and spoiler in the old car, but is concentrated on the wing in the new car. The NASCAR R&D Center measures about 1,100 pounds of downforce (570 pounds in front and 530 pounds in the rear), although they caution that they—unlike the race teams—aren't trying to get the maximum downforce possible.

While CFD calculations are a mainstay in the arsenal of most teams, they don't always tell the entire story. CFD results predicted that the new car wouldn't be as sensitive to aero push because the wake is not as turbulent. Calculations also indicated that passing should be easier because the low-pressure regions on the sides of the cars are wider, but the first things drivers complained about were how hard it is to pass in the new car and how bad the aero push is. That doesn't necessarily mean the simulations are wrong.

"A CFD simulation is a photograph," Eric said, "It's a snapshot of the situation. Five to ten years from now, we'll have a small movie clip."

Unfortunately, aero-movies won't come in time for the start of the 2008 season, when the new car meets its ultimate challenge: its first appearances on intermediate-size tracks, where downforce is critical. You can bet that race team aerodynamicists will be staying up late with their computers trying to figure out how to get that small improvement in downforce that will send their car to Victory Lane.

Eight

Texas Motor Speedway at 150 mph

Driving west on Texas Highway 114 from the Dallas–Fort Worth airport, you get the feeling after about twenty minutes that you aren't heading toward anything. The strip malls get farther and farther apart, but finally you reach the top of a hill and there it is, over the trees, like a blue and white version of the Emerald City: the Texas Motor Speedway. TMS is a 1.5-mile intermediate track that holds more than 200,000 people.

The track dwarfs the surrounding town. The woman at the hotel's front desk says they are "sort of on the border of civilization." I had to backtrack a couple of miles to find something other than fast food. It was a dreary, gray Friday in early February. Driving back again toward the track, a mist from the rain that had been falling much of the day enhanced the Emerald City effect.

I was up early the next morning, even though I didn't have to be at the track until 11:30, so I lingered over coffee and reviewed some numbers. At 150 mph, you move a little more than seven feet in the blink of an eye. In one second, you travel almost three quarters of a football field. Taking the 24-degree banked turns at TMS at 150 mph produces a centripetal acceleration of about 2 g. I understood the numbers in my brain, but today, I was going to gain more of a gut-level appreciation of them.

I headed over to the Speedway at about 11:15 and took the south tunnel under the track and into the infield. After looking over the cars lined up in the garage, I took a seat in one of the white plastic lawn chairs in a section marked DRIVERS ONLY. A whiteboard with a drawing of the track stood in front of the chairs, which filled quickly with about ninety people—including seven or eight women. Families and friends stood in the back, watching, giving words of encouragement, and taking pictures.

A little after noon, Mike Starr, a Texan with a wry grin, stood in front of the whiteboard to give us a brief tutorial. Mike, the owner of the Team Texas High Performance Driving School, explained that the Chevy Monte Carlos we were about to drive are former NASCAR cars, with NASCAR-standard SB2 engines. The engines are detuned by using a smaller camshaft and a smaller carburetor, so they "only" put out about 600 horsepower.

Despite being detuned, Mike warned, these are still probably the most powerful cars any of us will ever drive. He added that even the smartest person sometimes loses his brains when getting behind the wheel, and he asked us to please try to keep our brains with us at all times.

The instructions were minimal, which is good because everyone was getting a little anxious. Mike spent the most time on how to get around the track—and no, he didn't just tell us to turn left. The preferred line, Mike explained, would let us go faster and help the instructors know what to expect from the ten cars that would be on the track at a time.

Texas is a quad oval, which means that the track is basically an oval; however, the front stretch (the part of the track with the start/finish line) has two angled segments. The strange shape is why the frontstretch is 2,250 feet long and the backstretch is only 1,330 feet long. Dashed lines run along the straightaways and white boxes are painted near the turns: We are to let off the gas at the end of the dashed lines, have both wheels on the inside of the first box, hug the bottom of the track and maintain our speed until the second box, and then accelerate, swinging out to the lane closest to the wall coming out of turns 2 and 4.

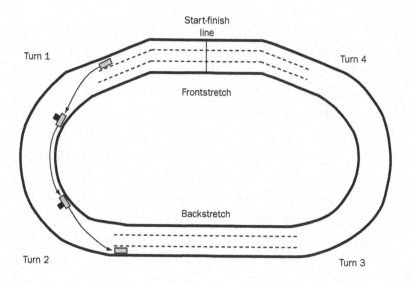

We climbed into passenger vans and traced the preferred line around the track. Although the speedometer read 80 mph, it didn't feel at all fast. The front stretch baffled me a bit, but our driver explained that it wasn't me.

"Look—I'm not moving the wheel," he said, "but we're not going straight either." It looked like the van was going straight and the track was wiggling from side to side—that's the effect of the 5 degrees of frontstretch banking and the quad-oval shape. After the van ride, we were divided into groups of ten. Seats have to fit the driver snugly, so we were matched with a car according to our sizes.

After donning a firesuit and finding an appropriately sized helmet,

I watched the cars in the first few groups circle the track. I was relieved to not be in the first group, because it gave me a chance to see that Mike was serious about going as fast—or as slow—as you feel comfortable. I ran through the instructions in my head: into the turn, off the gas, both wheels inside the box, hold speed to second box, accelerate, swing out to wall, and repeat.

The first challenge was getting into the car. Standing on the driver's side facing forward, I swung my right leg into the window, and then transferred my weight so that I was sitting on the windowsill. Professional drivers get in using one smooth motion, because sitting on the windowsill is not a comfortable position. The windows are about thirty inches by fifteen inches, which gave me just enough room to grab my left leg and pull it into the car without dislocating my hip. With both feet on the car floor, I could contort my way into the seat. I'm five-foot-six and very flexible. Six-foot-five Michael Waltrip is either one very limber guy or has a high tolerance for pain.

I felt a little claustrophobic in the car, especially after one of the employees slipped the steering wheel onto the column. The cockpit is so small that the steering wheel has to be removable to allow the driver to get in and out of the car. The seat, which enveloped me, was low in the car. The rib braces on the seat came up to my underarms and stuck out so far that there was no easy way for me to put my arms down.

A race car has three pedals—clutch, brake, and gas, left to right—just like any standard-transmission car. The clutch took a lot more force than I'm used to applying. I suspect this particular car used to be a road-course car, because the brake and the throttle were extremely close to each other. A piece of metal stuck up vertically from the right-hand side of the throttle to keep my foot from slipping off the pedal in the turns.

The most noticeable modification to the car was the addition of a second seat, which was occupied by my instructor. Paul, who I estimated to be in his early fifties, was relaxed and efficient as he reviewed the hand signals with me. Thumbs up means go faster, thumbs down means go slower, and a flat hand means level off. He instructed me to put my hands at eight and two because that makes it easier to turn left and, he told me, for him to reach under my hands to help me get the car going along the right line if necessary.

I expected to be nervous, but I wasn't, probably because I was preoccupied running through the instructions about where the car needed to be with respect to the white lines and boxes. The cars are push-started to give the transmissions a little longer lifetime. I started with the car in second and the clutch pressed down. Mike had warned us that the accelerator really revs the engine, so we should let the clutch out all the way before depressing the throttle. We trundled down pit road, and when Paul motioned, I pushed in the clutch and shifted to third. The shift lever is spring loaded and moves to the right all by itself.

I released the clutch and stepped on the gas. Actually, I stepped on the gas *and* the brake because they are so close to each other. Paul and I both shouted, "Whoa!" One disadvantage of a high-throughput driving school is that you feel like you're pulling out of a rental car lot without stopping to adjust the mirrors or check where the turn signals are—except you're going a lot faster. I recovered, shifted cleanly to fourth gear, and moved out onto the track.

Driving a stock car is nothing like driving a street car really fast. A race car is set up to pull to the left, so you sometimes have to steer right to go straight. I had no peripheral vision, and I was so focused on staying on the right line that I didn't even use the mirrors. The engine is optimized for speed, so when you're puttering along at a mere 100 mph, the engine chugs and huffs uncomfortably.

The solution, of course, is to go faster. Paul gave me the thumbs-up, letting me increase my speed on successive circuits. I realized that the place to start turning so that I would be under the white box was when I first saw the black streak on the wall of turn 3. As we completed laps, I started to appreciate what spotters mean when they tell their drivers, "Get back in your rhythm." The repetition of the slow-constant-fast sequence around the corners became comforting. I learned to use the seat to brace myself as we were thrown to the right around the corners.

As I passed the start-finish line on what I thought was a particularly good lap (Paul had given me the level-off sign on the backstretch), I realized the flagman had just waved the white flag, signaling that this would be the last lap. Mike had warned us to resist the urge to think, "This is the last lap, now I need to go really fast," but before I even finished that thought, we were through turn 4 again and the checkered flag was waving. Paul switched off the engine as we came out of turn 2 and we coasted into the pits.

It didn't matter that the cars don't have speedometers, because I don't think I took my eyes off the track once. I asked Paul how fast we went. I figured we maybe hit 120 mph, so I was surprised when he told me that when he was giving me the level-off sign, we were at the maximum speed of 148 to 150 mph.

"We passed a lot of cars out there," he said approvingly.

It wasn't until I climbed out of the car and started comparing notes with my fellow drivers that I realized my legs were wobbly and my heart was racing. I learned later that it was adrenaline. Professional drivers shake the same way after their qualifying runs. Even though my brain didn't realize we were going fast, the rest of my body did.

The day wasn't over yet. I had signed up to ride a few laps in a car driven by one of the instructors. I had done this for two reasons: First, I was concerned I might be too scared to go very-fast driving myself, and second, I wanted to be able to focus on the car's motion without worrying about hitting anything.

I rode with Mike, who led a pack of ten cars onto the track. Mike swerved the steering wheel back and forth to warm up the tires. It looks like a gentle motion on TV, but it doesn't feel that way inside the car. We moved up onto the track, Mike punched the throttle and we started to accelerate.

Acceleration is how fast your speed changes. Coming onto the track, I estimated that we went from about 70 mph to 140 mph in around five seconds, which means we increased our speed on average by 14 mph each second. We started at 70 mph. After one second we were going 84 mph, after two seconds were going 98 mph, and so on.

A "g" is the acceleration due to gravity. When you drop something, its speed increases by 32 feet per second (22 mph), every second. Our acceleration getting onto the track was about 0.6 g, which is the same acceleration as going from 0 to 70 mph in five seconds. During a race, a NASCAR car can accelerate from 0 to 200 mph in 12 seconds, which is 0.76 g.

The force you feel accelerating at 0.6 g is 0.6 times your weight. A 160-pound person undergoing an acceleration of 0.6 g experiences a force of 96 pounds. Taking off on the space shuttle creates an acceleration of 3 g, which feels like three people your weight sitting on top of you. A 3,600-pound car accelerating at 0.6 g experiences a force of a little more than a ton. The ton of force making the car accelerate is above and beyond what the engine has to generate to overcome air resistance and friction between the tires and the track.

Although the engine propelled the car forward, I felt like I was being pushed backward when we accelerated. The first part of Newton's Law of Inertia says that a body at rest remains at rest unless something makes it move. I was moving because I was strapped tightly into the car's seat. The track exerted a force on the tires, the tires exerted a force on the chassis, the chassis exerted a force on the seat, and the seat exerted a force on me. The force I felt was the seat going forward while my body was trying to remain at rest.

The second part of Newton's Law of Inertia (as applied to cars) says that a car moving at 150 mph keeps moving at 150 mph in the same direction unless something causes it to change its speed or direction. When we braked (which didn't happen until the very end of the four-lap run), my body tried to keep moving forward. The brakes exerted a force on the wheels and so on, until the seat and its attached belts exerted a force on me that kept me from continuing forward at the same speed and in the same direction.

As we started into the corner, I noted that Mike didn't get off the gas at the end of the dashed lines on the straightaway. He was still on the gas and stayed on the gas throughout most of the corner, which made my head feel like it was being pulled out the window. I had to press my head back against the seat to keep it steady. The combination of higher speeds in the turns and the lack of rib braces on the seat enhanced the feeling of being pushed outward.

Newton's Law of Inertia talks about speed *and* direction. If you're going the same speed but changing direction, something—a force— is making you change direction. You may have heard this described as "centrifugal force," but there isn't really a force pushing you out- ward.

Get a soccer ball or basketball and a yardstick. Roll the ball in a straight line and hit it with the yardstick to make it roll in a circle. The ball wants to go straight, so to make it turn, you have to hit it at an angle with respect to the direction it is trying to move. Making the ball move in a circle requires you to hit the ball repeatedly, and you always have to hit toward the center of the circle. This inward-pointing force is called the centripetal (which means "center-seeking") force. The centripetal force is the reason cars turn. Just as you feel pushed backward when you accelerate (even though the force is pushing you forward), you feel pushed toward the outside of the circle even though the force making you turn is directed toward the center of the circle.

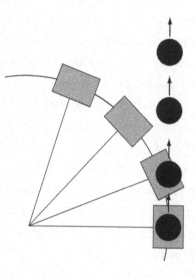

Jimmie Johnson experimented with Newton's Law of Inertia during the 2006 off-season. A golf cart turns because the road exerts a force on its tires. If you're seated in (or holding onto) the golf cart, the cart exerts a force on you that makes you turn, too.

The corollary—as Jimmie discovered—is that if you're not holding onto the cart, the cart turns (or stops) without you. You keep going in the same direction you had been heading. In addition to showing that Newton's Law of Inertia worked, Jimmie also learned how the strong the bones in his wrist were. Or weren't.

The acceleration you feel around a turn is equal to the speed squared divided by the turn radius. The size of the force needed to turn is your mass times this acceleration. You need more force to turn in a smaller radius or at a higher speed. If you double your speed around a corner, you need four times as much force to make the turn.

NASCAR drivers routinely experience 2 to 3 g in the turns, while Indy and Formula One drivers can experience 4 to 5 g accelerations because their cars travel at higher speeds and make tighter corners. Large accelerations can confuse the inner ear (which controls your balance) and produce disorientation, dizziness, and nausea. The inner ear's disequilibrium is why some carnival rides make people sick.

Sustained large accelerations can have more serious physical effects, including temporary blindness and blackouts. The Firestone Firehawk 600 CART race, which was to be held at the Texas Motor Speedway in 2001, was cancelled when twenty-one out of the twenty-five drivers reported disorientation, dizziness, nausea, and vision problems during practices. The speeds of 220 to 250 mph produced accelerations of more than 5 g that lasted about six and a half seconds in the turns, giving the drivers only four to six seconds on the straightaways before turning again.

At the 150 mph I estimate we were going around the corners while Mike was driving, the centripetal acceleration was right around 2 g, but no equation can describe how much more exciting 2 g is with another car less than a foot from your window.

As I exited TMS and headed back down Highway 114 to get some food, I counted no fewer than five cars pulled over by police. I made a mental note to watch the speedometer on my way back to the hotel.

While waiting for my food, I replayed in my mind Paul's statement that I "did pretty good for a first time" and started to wonder about whether I should look for sponsorship. If I can climb in a stock car with no previous experience and hit 150 mph after ten laps, how hard could it be to work up to 200 mph? I started scratching out a calculation on my napkin.

Races are won and lost in the turns. How fast you can get around the corners is determined by how much turning force you can generate from the friction between the tires and the track. Here are the numbers: TMS has a turn radius of about 750 feet. A 150-pound driver makes a car-plus-driver weight of 3,600 pounds. Cornering at 65 mph requires a turning force of 1,355 pounds, which corresponds to an acceleration of 0.38 g. Going around the same corner twice as fast requires a turning force of 5,419 pounds—an acceleration of about 1.5 g. Assuming equal weight distribution (to make the calculation easier), each tire has to generate about 340 pounds of turning force at 65 mph but 1,355 pounds of turning force to take the same corner at 130 mph, and a whopping 2,317 pounds—more than a ton—at 170 mph. If you can only get 1,100 pounds of turning force from each tire and you're headed into turn 1 at 130 mph, you have two options: slow down, or hit the wall.

The turning force generated by the tires depends on how hard the tires are pushed into the track, and how well the tires stick to the track. In the simplest case, the force pushing the tire and the track together is the weight supported by that wheel. If we assume that each wheel supports the same weight, our 3,600-pound car and driver would provide 900 pounds of force pushing each tire into the track.

A street tire on asphalt generates a turning force of about 80 percent of the force pushing down on it. If we put street tires on our race car, each tire would generate about 720 pounds of turning force. Racing tires are stickier, so they have more grip—about 120 percent of the force pushing down is a reasonable number, so our car with racing tires could generate 1,080 pounds of force per wheel. With that amount of force, we can go a maximum of about 116 mph around the turns.

Even I went faster than that around the turns at TMS, which made me realize that I was missing something. I hadn't factored in the banking. A track pushes on the car perpendicular to the track surface. On a flat track, the push is straight up and exactly offset by the car's weight pushing down. When a car turns on a flat track, all of the turning force has to come from the tires.

A banked track also pushes perpendicular to the surface, but because the surface is tilted, only part of the force pushes straight up. The rest of the force pushes toward the center of the turn, which adds to the turning force generated by the tires. There is a slight penalty because not all of the frictional force on a banked track points toward the center (some is keeping the car from sliding up the track); however, the increased turning power due to the banking more than makes up for the 20-percent loss in frictional turning force relative to a flat track. TMS has 24-degree banking in the corners, which means that you could take the turns at TMS at about 70 mph with absolutely no friction. Assuming the car generates a turning force equal to 100 percent of the force pressing down on that tire, the maximum speed around the corners at TMS is about 170 mph for a stock car. With the same assumptions, you could take the turns at Talladega (which have a larger radius and 33-degree banking) at more than 270 mph.

Banking at NASCAR tracks ranges from almost nonexistent (the 6-degree banking of turn 3 at Pocono) to extreme. Talladega's 33-degree banking means that the track rises twenty-six feet from the inside of the forty-eight-foot-wide track to the outside. The height increases by six and a half inches for every foot you walk up the track. The more moderate 14-degree banking of Fontana has about a three-inch increase in height for every foot along the track. I've illustrated the comparison of the banking at the two tracks on the next page.

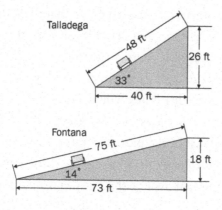

The banking at TMS provides 2,848 pounds of turning force at 170 mph. Cornering at 170 mph requires a total of 9,267 pounds of turning force, so the 6,419 pounds of force not provided by the banking has to come from the tires. If the force is divided equally between the tires, we need 1,605 pounds of turning force per tire.

The car and driver only provide 900 pounds of force downward on each wheel, which is the mechanical grip. The rest of the grip comes from aerodynamic downforce. Downforce makes our analysis more complicated because the amount of downforce changes depending on how fast you're going. At TMS, a car might enter a turn at 180 mph but slow to 150 mph in the middle of the turn. The 2,000 pounds of downforce you have at 180 mph decreases to 1,389 pounds of downforce at 150 mph.

Changing downforce means changing grip. I put sandbags in the pickup bed in the winter because the extra weight gives me more rear mechanical grip. You can think of aerodynamic downforce as a bag of sand that changes weight depending on how fast you're driving.

There is no free lunch in racing—or in physics. Although turning force increases as the force pushing down increases, that only works up to a point. After that, you actually get *less* increase in turning force as you push down harder on the tire.

There's more bad news. For simplicity, I've assumed that the forces on the car are distributed equally across all four tires. If that were true, all four wheels would lose grip at the same time. I put sandbags in the back of the pickup because when the bed is empty and it's icy (which is the same thing as a race car losing grip), having less weight in the rear makes the back end of the truck want to swing out when I turn. This is what drivers call "loose"—the rear tires have less grip than the front tires. If the front tires have less grip than the rear tires, the car is "tight" and doesn't want to turn. (Open-wheel drivers use "understeer" and "oversteer" for "tight" and "loose.") Tight means you see the crash coming and loose means you don't, because a loose car usually hits the wall tail-first.

Crew chiefs rejoice when their driver tells them the car is balanced, meaning neither loose nor tight. A balanced car means that the wheels are equally loaded. You can adjust the weight before the race so that each tire supports the same amount, but weight shifts as the car moves.

When you brake, the car's weight shifts forward, putting more weight on the front wheels than on the back wheels, which gives the front wheels more grip than the rear wheels. The opposite happens when the car accelerates. The weight shifts toward the rear of the car, giving you more rear grip than front grip. When you corner, weight shifts from the inside wheels (usually the left side) to the outside (right-side) wheels.

The amount of weight that shifts depends on the car's acceleration, the distances between the wheels, and the height of the car's center of gravity. The center of gravity (CG for short) is a unique spot on the car. The CG is the one point on the car where it balances perfectly. The CG of the old car was about a foot off the ground while it was in motion. The new car's CG is about 2½" higher, which allows more weight to shift on braking, acceleration, and cornering, and produces more change in the amount of grip each wheel has. Elliott Sadler compared moving from the old car to the new one to stepping out of your sports car into a big SUV.

When you turn left and accelerate, as you do coming out of a turn, weight shifts from front to back and from left to right. The right rear wheel supports more of the car's weight and the left front supports the least, which means that the right rear tire has much more grip than the left front tire. The left-front wheel of an ill-handling race car can actually lose contact with the track coming out of a corner.

This brief analysis just begins to answer the question of why driving a race car is very different than just driving your own car really fast. The driver needs to get as much turning force from the tires as possible. The turning force is determined by the grip, but each wheel has a different amount of grip and that amount constantly changes, depending on your speed and whether you're accelerating, braking, or turning. The driver essentially has a different car at each point around the track, and that's without taking tire wear or changes in track temperature into consideration.

As my food arrived, I put my scribble-covered napkin to one side and concluded that the difference between Jeff Gordon and me is the difference between an artist and someone who paints. I'm not giving up, though—even Van Gogh had to start somewhere—and (listen up, potential sponsors!) I'm a pretty quick learner.

Nine

Sound Thinking

Atlanta should be warm, but the mid-March Friday morning I arrived, it was decidedly cold and wet. Those of us standing in line at the NASCAR credentials hauler outside the Atlanta Motor Speedway were catching the tail end of a line of storms that had started the previous afternoon. My rain poncho wasn't quite enough protection from the downpour, and I very much appreciated the young man from Motor Racing Outreach who shared his umbrella with me while we queued up.

I didn't mind the wait. Try asking a major league baseball team to let you wander around the field—much less the locker room—before a game. NASCAR is a surprisingly accessible sport for fans, but they do have to keep track of who is going where.

I met Josh Browne, team director for the Gillett Evernham Motorsports Dodge Dealers/United Auto Workers Dodge at a Society of Automotive Engineers motorsports conference. Josh is serious about racing, but he has a wonderfully wicked sense of humor. When I asked what variables he uses to decide on adjustments during a race, I expected to hear about lap times and tire wear. Josh thought for a moment.

"How loud the driver is yelling at me. How many swear words he's using. Stuff like that."

I asked Josh if I might embed myself with his team during a race weekend to better understand what happens when the car gets out on the track. Since both the old and new versions of the car were being run in 2007, he suggested Fontana (old car) and Martinsville (new car). After getting all the necessary clearances, we were set.

Or not. Atlanta (where I was shivering in the rain) is neither Fontana nor Martinsville. Although Atlanta was the fourth race of the season, it was only Josh's second race due to "Boltgate." Josh was suspended for the first two races of the season for using spoiler bolts with holes in them. NASCAR said the bolts provided an unfair aerodynamic advantage. GEM argued they had used the same bolts in 2006 without any issue, but they lost their appeal.

That's why I was squishing my way across the still-wet grass to the tunnel that runs under turns 3 and 4 and into the infield of the Atlanta Motor Speedway. Flashing my bright orange garage pass at the officials, I passed through the wrought iron fence that separates the garage from the rest of the infield and started looking for my hosts.

Each team's hauler enters the garage and lines up in order of owner's points, with the highest-ranking cars closest to pit road. It took me a while to find the No. 19 hauler because I started by looking in the wrong place. Although the team was tenth in points, NASCAR uses the previous year's points for the first five races each season. The team ended 2006 in thirty-fourth place after a tumultuous year in which driver Jeremy Mayfield was replaced for the last fourteen races by Elliott Sadler. I finally found the No. 19 hauler between the No. 21 Little Debbie's hauler and Kyle Petty's No. 45 Marathon hauler.

A hauler is a two-story semi-truck designed to transport everything the team needs from the shop to the track. A motorized lift raises the cars to the second story, which is just large enough for two cars (the primary and a backup) to fit nose-to-tail. The bottom level, which is about six and a half feet high, has a narrow corridor through which two people can pass if one is turned sideways. The cabinets lining the inside of the hauler contain everything from shocks, springs, and an extra engine to food, radios, and firesuits.

The gray hauler interior is accented with splashes of red, white, and black. A wide red band atop a narrower black band runs along both sides of the hauler about a foot down from the ceiling. There are three black countertops: The shock specialist's station is in the back on the left-hand side as you enter, and there's a general-purpose prep area halfway back on the right. The counter closest to the front, on the left, is for some of the most important racing equipment: food and coffee.

Everything is overseen expertly by Chris Miko, a Bronx native who is hauler driver, pit-crew member, and den mother. In addition to getting the hauler to the track, Chris is in charge of details like making sure the hauler is stocked with the crew members' favorite foods, charging the radios, and making the area just outside the hauler appropriately welcoming with folding director's chairs and a cooler full of soda and water.

The No. 19 is a sister car to Kasey Kahne's No. 9 car. They have the same sponsor, but their paint schemes are inverses of each other. The No. 9 car is white in the front and graduates to red in the back, while the No. 19 is red in the front and white in the back. Josh and engineer Chad Johnston, who are squirreled away in the office at the very rear of the hauler, are a similar study in complementary contrasts. Josh, a thirty-six-year-old from Philadelphia, is brown-haired and boisterous, with glasses and a prominent chin. Chad is ten years younger, blond, shorter, and quieter. Chad has an engineering degree from Indiana State University, while Josh's degrees are from Carnegie Mellon and Oakland University. This is Chad's first year on the road with the No. 19 team, but he and Josh already have established a relationship that is a cross between fraternity brothers and an old married couple: They bicker, they crack jokes, and they finish each other's sentences.

The haulers park across from the garage, forming a thruway to pit road. Atlanta is a mile-and-a-half track with a huge infield. In addition to the garage, there are campgrounds for fans and competitors. You'll find everything from people sleeping in tents to RVs that cost more than most people's houses. The latter mostly belong to the drivers, but when you spend thirty-eight weekends a year at tracks, those RVs essentially *are* their houses.

Crew members' travel schedules are easily as demanding as the driver's (if not more so), but you won't find them in the RV lot or at fancy hotels. They tend to stay two to a room, usually at the types of hotels that offer free breakfast. As long as the hotel has a comfortable bed, Chris told me, it doesn't matter because they spend almost all of their time at the track.

An engine in the garage started up and Chris laughed when I jumped. I knew the cars were loud, but they're really, really loud when they're twenty feet away. I hadn't expected to need my earplugs that early in the morning.

Sound is made of pressure waves. Speaking makes your vocal cords vibrate, causing the air molecules near the vocal cords to oscillate back and forth. How fast the air molecules move back and forth is determined by the frequency of your voice, which in turn is determined by how fast your vocal cords vibrate. Higher vibration rates produce higher frequency sounds.

The oscillating air molecules bump into nearby air molecules and start those molecules oscillating. A sound wave moves like The Wave in a stadium. Each person's up-and-down movement starts when the person next to them moves. Although everyone stays in their original seats, the wave travels all the way around the stadium. In the same way, the air molecules closest to my mouth don't travel to your ears; they just bump into other air molecules and eventually the air molecules near your ears start vibrating.

Your bowl-shaped outer ear helps you collect and focus sound waves. High-pressure air makes your eardrum—a tightly stretched membrane that separates the outer and middle ear—move inward, while low-pressure air pulls it outward. Variations in the pressure make your eardrums vibrate in the same pattern as the sound wave. The hammer, anvil, and stirrup bones, which are connected to the eardrum, amplify the vibration. The vibration finally reaches the cochlea, where more than 20,000 hairlike receptor cells (which are called "hair cells" even though they are in your ear) of different lengths convert vibrations into electrical impulses that are interpreted by your brain as sound. Humans can hear sounds from 20 hertz to 20,000 hertz. Twenty hertz means that the wave oscillates back and forth twenty times each second. We hear best from a few hundred to a few thousand hertz.

Loudness is determined by the size of the air-molecule oscillations. A very loud sound corresponds to very high and very low pressures that can make the eardrum try to flex more than it is capable of. These large-amplitude pressure waves can literally break your eardrum. Of more concern are very loud noises that can permanently damage your hair cells. Few animals are capable of producing new hair cells. Sharks and chickens can, but people can't. A damaged hair cell doesn't get replaced, and you can't hear without hair cells.

Loudness is measured in decibels (dB). The decibel scale is a logarithmic scale: Every 10 dB increase corresponds to a multiplication of the loudness by 10. Zero dB is the faintest sound a human ear can detect. A 20 dB sound (a whisper) is ten times louder than a 10 dB sound and a hundred times louder than a 0 dB sound. A 60 dB sound, which corresponds to normal conversation, is one million times louder than a 0 dB sound. A NASCAR engine is around 120 dB, which is 1,000,000,000,000 times louder than a 0 dB sound. The scale is logarithmic because that is closer to the way human hearing works: One sound may be ten times more intense than another, but the human ear perceives it to be as twice as loud.

The louder the sound, the less time you can safely listen to it. The Occupational Safety and Health Administration recommends limiting exposure to 90 dB sounds to no longer than eight hours and 115 dB sounds to no longer than fifteen minutes. Since the sound level is 120 to 140 dB near the car (and 110 to 125 dB inside it), I had brought molded rubber earplugs. I also had the headphones from my scanner, which fit over my ears. Both have a noise reduction ratio (NRR) of 30 dB, which means they reduce noises 30 dB. Either one would make 120 dB sound like "only" 90 dB. That's a thousand times lower, which could be the difference between permanent hearing loss or not.

Race cars are loud primarily because they don't have mufflers. Sound waves, like all waves, can interfere with each other. A "dead spot" is a place where the maximum of one wave overlaps the minimum of another wave. High-pressure and low-pressure regions mix, reducing the sound's loudness.

A muffler uses sound against itself. Sound waves are carried by exhaust gases into the muffler. The sound waves reflect from the inside of the muffler and interfere with each other, decreasing the loudness. Engines generate many frequencies of sound. The muffler length determines which frequencies are muffled best. It is impossible to cancel all frequencies with a single-length muffler, so mufflers are designed to work most effectively at the frequencies the engine produces most loudly. Race cars don't have mufflers because the buildup of exhaust gases would create a backpressure. The resulting traffic jam for the gases trying to get out of the engine would limit how fast the engine could run.

Two-way radios allow the team to communicate with the driver while he's on the track. Once practice starts, it's too loud for team members to talk directly with each other, so they use the team radio to communicate. A microphone converts the pressure of the moving air molecules from your voice into an electrical signal. The electrical signal travels through an antenna, where it is converted into an electromagnetic wave. Most of the NASCAR team frequencies are around 400 to 500 megahertz (MHz), which means that the electromagnetic waves oscillate 400 to 500 million times each second. When the electromagnetic waves reach the listener's radio receiver, the antenna converts the electromagnetic wave into an electrical signal, and the earpiece converts the electrical signal back into a pressure wave that the ear then detects.

The most important advantage of converting sound waves into electromagnetic waves is speed. Sound waves move about 770 mph. Electromagnetic waves move at the speed of light, which is 671,000,000 mph. Electromagnetic waves also don't dissipate as quickly as sound waves, so they can travel farther.

Before radios, team members shouted at each other and communicated with the driver by holding up pit boards—large signs on which they'd write lap times or instructions like PIT NOW. The driver used visual signals, like tapping on the roof for "loose" or on the door for "tight." Each team now has its own radio frequencies, which allow them to communicate with each other. Fans can also listen in to their favorites team's communications using a scanner. When you tune to your team's frequency, you're telling your scanner which electromagnetic waves to detect.

I was warned by a veteran race-goer to always start by adjusting my scanner to the minimum volume necessary to hear. Your ears adjust to noise easily. If you start with the volume too high, you won't realize that it's too high after very long. More people in their twenties and thirties have serious hearing loss than ever before, which doctors attribute to earphones and excessive volume. If other people can hear your scanner (or iPod), you're probably causing yourself permanent hearing damage.

Chris has about twenty-five radios in the hauler, which is enough for the crew and a few guests. It takes a lot of people to get a car ready to race. Fifty cars would try to qualify for the forty-three spots in this weekend's race. There wasn't enough garage space, so the teams at the bottom of the owner's points list worked outside the garage under large canopies. In addition to drivers, crew chiefs, and mechanics, there are journalists and television reporters, public relations people, manufacturer's representatives, and fans. Lots of fans.

"You think it's busy today," Chris said knowingly, "wait until Sunday."

Teams are used to having fans in the garage. Chris had a metal stand full of "hero cards"—glossy cards with pictures of driver Elliott Sadler—in front of the hauler for fans to take with them. The team appreciates the fans' enthusiasm and dedication; however, having so many people around the garage sometimes makes it difficult to get your work done, as tire specialist Scott "Swifty" Swift explained.

"I'm down like this," he demonstrated, kneeling as if taking the cap off a tire stem, "and this woman comes rushing over because she sees Tony Stewart. And she knocks me over trying to get to him. Tony, he comes over and helps me up, then he tells the woman, 'Lady, I can't tell you how happy it makes me that you getting my autograph is more important than this guy doing his job.'"

Tony did sign an autograph for the woman, but if you are fortunate enough to get a garage pass, keep in mind that you're visiting an office during working hours and that everyone working there has an impending deadline. It's not hard to be a good garage visitor. There are only a few rules, and they are mostly common sense.

Watch where you're going. Anything with wheels or wearing a uniform has the right of way. If crew members are there, they're working—even if it doesn't look like it to you. Don't make someone ask you to move more than once. Never lean on a stack of tires that don't belong to you. Only take things that are clearly for fans, and then don't take more than one. Don't go places you haven't been invited.

And never knock over crew members, even if Tony Stewart is nowhere to be seen.

Ten

The Rubber Hits the Road

Every team has one person at the track dedicated to making sure the tires are ready to do their jobs. Scott Swift, who's called "Swifty," has been a tire specialist for twelve years.

"I started volunteering at a (race) shop," he said, "and I kept showing up until they started paying me enough that I could quit my other job." Swifty has worked with MB2 Motorsports, Haas Racing, and Joe Gibbs Racing. This was his second year with the No. 19 team.

Swifty knows a lot of people in the garage and, since he normally works just outside the hauler, is always within shouting distance of a joke. The only North Carolina native on the No. 19 garage crew, the thirty-eight-year-old has blond hair and a beard just starting to show flecks of gray. His sturdy build is useful for moving the sixty-pound wheel/tire combination.

Tires generate the turning force that keeps the car headed for the frontstretch instead of the wall. When a car turns, the front tires momentarily point in a slightly different direction than the wheels. The angle between those two directions is called the "slip angle," which is illustrated on the next page (left). The turning force the tire generates depends on its slip angle, as the graph on the next page (right) shows.

At small slip angles, the turning force is due mostly to the tires resisting being twisted by the turn. As the slip angle increases, part of the tire starts sliding. Larger slip angles mean more sliding. There has to be *some* sliding to get the maximum turning force, but too much sliding decreases the turning force.

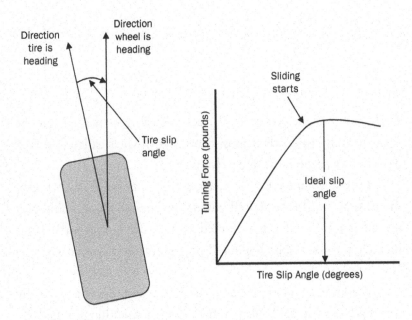

Tire slip angles are small—between two and four degrees. If a driver keeps the tire at a slip angle below the ideal value, he's not getting as much turning force as he could. If he drives with the tires above the ideal slip angle, the turning force decreases, and the car slides instead of turns. It's like *The Price Is Right*: Get as close as possible to the ideal slip angle without going over. The penalty for going over is a little steeper in NASCAR than it is on a television game show.

The first thing Swifty did when he arrived at the Atlanta Motor Speedway this morning was to pick up the team's Friday and Saturday tire ration from the Goodyear tent. NASCAR limits teams to six sets of tires for qualifying and practice. At impound races, where the team is prohibited from working on the cars after qualifying, they only get four sets. Swifty usually gets between ten and fifteen additional sets for the race, depending on the track and the length of the race. Each tire has a bar code that tells him when the tire was made. Swifty also can download additional information about each tire to his laptop from a database Goodyear makes available to the teams.

Every team gets the same tires, but not all tires are the same. At all races except the two road courses, left-side tires have a slightly smaller circumference (the distance around the tire) than right-side tires: 87.4 inches vs. 88.6 inches for Atlanta. The difference in circumference (called stagger) helps the car corner. The outer tires have to travel farther than the inner tires when the car turns. The larger right-side tire helps compensate for the longer distance.

Even though all the right-side tires have the same tire code (Goodyear's way of naming tires), there are minor differences between tires. Not all right-side tires are exactly 88.6 inches in circumference. Swifty measures the circumference of each tire, examines the information from Goodyear, and groups the tires into sets of four that are as similar as possible, down to the date and shift on which they were made.

Goodyear has supplied tires to NASCAR since the 1950s and has been the sole tire provider since 1994. Each tire costs $417, which does include mounting and balancing. This weekend's tire bill for the No. 19 team will be more than $30,000.

The tires stacked in front of the hauler look different than street tires because racing tires have a totally smooth tread; however, the basic construction is similar. A rubber-coated bundle of steel wires, called the bead, runs along both inside edges of the tire and seals against the wheel. An inside layer of rubber (the liner) holds gas inside the tire. The tire body is made from layers of rubber-coated fabric called plies. The next layer is the belt package, which provides strength and resistance to puncture, and the outermost layer is the tread.

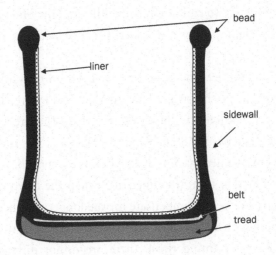

Goodyear has the unenviable challenge of creating tires that are "just right": soft enough to provide grip, but not so soft that they wear out too quickly. "Just right" is different for each track. In 2005, Goodyear used twenty different tire codes. That number jumped to thirty in 2007, when they had to have tires for the new car and the old car.

One of the most important factors in making the tires "just right" is the tire compound (or recipe) used for the tread. Saying that the tread is made of rubber is like saying that chocolate chip cookies are made of flour, sugar, eggs, butter, and chocolate chips. Like cookies, each rubber recipe has its own properties.

Rubber was the hot new material of the 1820s, but natural rubber, which comes from the sap of the rubber tree, is more like chewing gum than tires. Originally used for erasers, the rubber available then was unsuitable for almost anything else. Hot weather turned it smelly and limp, while it became brittle and cracked in cold weather.

The initial rubber boom was over by the 1830s for most people, but not for a frequently bankrupt businessman named Charles Goodyear. Goodyear was obsessed with finding a way to make rubber stronger and less sticky without losing its elasticity. His single-mindedness and previous failures made people a little skeptical of his efforts.

The story has it that as Goodyear was showing off his latest result to a snickering group at the Woburn, Massachusetts, general store in 1839 (he had found that adding sulfur to rubber made it marginally better), the piece of rubber he had been gesturing with flew into the air and landed on a stove. Goodyear expected to be cleaning up a gooey mess; however, the material on the stove wasn't melted. It was elastic, durable, and—unlike his previous attempts—didn't stick to everything.

Goodyear experimented some more and eventually settled on using steam to cure the sulfur-doped rubber. The curing process was called vulcanization after Vulcan, the Roman god of fire and volcanoes. Sadly, Goodyear never profited from his discovery. The Goodyear Tire & Rubber Company was named in recognition of his discovery, but Goodyear himself never worked there.

Goodyear was more interested in *how* to transform rubber than he was in *why* the vulcanization process worked. Rubber is a polymer called polyisoprene, which naturally comes in short chains with little "arms" located along a carbon backbone, as shown in the top illustration on the next page. These chains are not very strongly connected to each other, so natural rubber acts more like a fluid than a solid, especially when warm.

Vulcanization makes the chains longer and forms sulfur-atom bridges between them. These crosslinks give rubber its strength and durability, while still allowing it to be elastic; however, the crosslinks don't form until the rubber is heated.

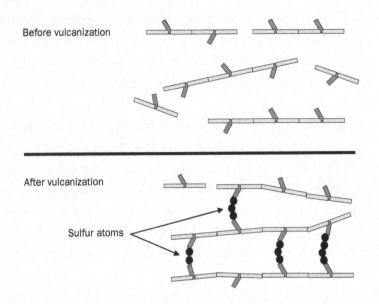

Before vulcanization

After vulcanization

Sulfur atoms

There are alternate ways to vulcanize rubber. As early as 1600 B.C., Mesoamerican peoples (the Olmecs and Aztecs) discovered that the juice of a local morning-glory vine vulcanized rubber-tree sap well enough to make balls used for sports and religious rituals.

Gas molecules escape through natural rubber rather easily, so most tires combine natural rubber with synthetic rubbers that are better at keeping gas molecules inside the tire. Racing tires are about 65 percent synthetic and 35 percent natural rubber, while passenger car tires are about 55 percent synthetic and 45 percent natural rubber.

Rubber has very good frictional properties. When two metals rub, microscopic protrusions scrape along each other. Rubber, in contrast, deforms to wrap around protrusions and fill dips, which makes rubber ideal for producing grip. Goodyear develops different tire compounds for different races, taking into account track banking, speeds, and the surface type. They can change the proportion of natural to synthetic rubber, the specific elements used, the sidewall structure, the mold use to shape the tire, and the vulcanization process. Each of these factors changes the way the tire grabs the track.

Finding the right balance of properties is difficult. More friction means more grip, but also more heat and more wear. At high temperatures, a very thin layer of the tread melts. The track scrapes off some of the melted layer and it mixes with dust and other debris. These deposits are called "marbles," not because they are round, but because driving over them feels like driving over marbles.

Heat dissipation is the reason racing tires don't have tread patterns like passenger-car tires. The flat surface of a racing tire (called a slick) helps dissipate heat better. The thinner tread on a racing tire—about ⅛" compared to ⅜" on a new passenger-car tire—also improves heat dissipation.

The grooved treads on passenger-car tires are primarily to prevent hydroplaning, which happens when water gets between the tire and the road. The car loses grip because there is little friction between the tire and the water. Deep grooves in the tire give water an escape path so that more of the tire maintains contact with the road. NASCAR doesn't race when it is wet, so hydroplaning isn't a concern.

Right-side tires support more of the car's weight during cornering and thus wear faster. Left-side tires for oval tracks are made from a softer compound so that both sides wear at about the same rate. Softer tires are used on flatter tracks, which generally have lower speeds. Atlanta is a rough asphalt track with some of the highest speeds on the circuit, so tire wear will be a concern.

Each tire has a row of five small holes called "wear pins" running across its tread. The wear pins are diagonal across the tread on the right-side tires and straight across on the left-side tires. Swifty uses a wear gauge to measure the depth of each hole relative to the tire surface and records those values on a sheet attached to his metal clipboard.

As soon as the tires are removed after the first practice run, Swifty will heat the tire and the end of a five-in-one tool with a propane torch and scrape away any excess rubber from the wear indicators. He'll compare the depth of each wear indicator to its starting value, which tells him how much tread has been worn away. A tire can lose 15 to 20 mils (thousandths of an inch) after ten laps—and remember that it starts with only 125 mils or so of tread. Swifty measures the wear in five separate spots to determine whether the force is distributed evenly across the tire. He measures and records the tire temperatures because heat—which is generated by friction—also indicates where the tire has the most grip.

Goodyear has the challenge of bringing "just right" tires to thirty-eight races at twenty-one different tracks. The challenge is magnified when tracks are resurfaced or redesigned. The May 2005 race at Lowe's Motor Speedway was the debut of a new, partially levigated track surface. Levigation is a fancy word for diamond grinding, which had been used to smooth out bumps in the racing surface. The 600-mile race had a record twenty-two cautions. President H.A. "Humpy" Wheeler decided that only a full levigation would solve the problem and ordered it completed before the fall race.

Teams arrived in October hoping they wouldn't have the same problems they had in the spring race. They didn't: The problems were worse. The 500-mile race had fifteen cautions, eleven of which were caused by tire problems. Tony Stewart wondered aloud on the radio whether his life insurance was paid up.

Levigation smoothed out the large bumps, but the grinding made the track surface rougher. The rough surface generated more grip, so the cars ran faster and the tires got hotter. The tire compound couldn't dissipate the heat fast enough, leading to blistered tires and blowouts. Teams complained that Goodyear should have brought a harder tire. Goodyear complained that teams ignored the recommended minimum pressure, which is directly related to how much force the tire can withstand.

A tire isn't round when it's supporting the weight of the car: The flat spot where the tire touches the ground is called the contact patch. The contact patch for a tire on my pickup is about four inches long by seven inches wide, which makes a contact patch area of twenty-eight square inches. The contact patch of a Goodyear racing tire is roughly thirty-six square inches, which is about the size of a man's size-11 standard-width shoe.

A typical tire pressure (for a consumer vehicle) of 30 pounds per square inch (psi) means that one square inch of the contact patch supports 30 pounds of the car. A contact patch with an area of twenty-eight square inches should support 28×30 pounds, or 840 pounds. All four tires on my pickup support 4×840 pounds, or 3,360 pounds, which is about the weight of the truck. If your vehicle calls for different pressures in the front and rear tires, calculate the weight supported by each tire separately and then add them together.

At Atlanta, Goodyear's minimum recommended tire pressures were 22 psi (left front), 20 psi (left rear), 48 psi (right front) and 45 psi (right rear). At Martinsville, which is a much shorter track with less banking and lower speeds, the minimum pressures would be 10 psi (left front and left rear), 23 psi (right front) and 22 psi (right rear). The right-side tires carry much more of the weight during cornering, which is why they are inflated to higher pressures.

The tread width on Goodyear Eagle racing tires ranges from 10.6" to 11.6", compared to about 8" for a passenger-car tire. A wider tire does not create a larger contact patch. If the weight of the car and the tire pressure remain the same, the contact patch will have the same area, regardless of the tires. What does change is the *shape* of the contact patch, and that affects how the tire slips.

A wide tire's contact patch is shorter and broader compared to a narrow tire, even though both contact patches have the same area. A tire starts slipping the same distance from the front edge of its contact patch, regardless of contact-patch shape. Compare the two contact patches in the illustration below when they start to slip. The wider tire has a much smaller area slipping, which is why the wider tire has more grip. The size of the contact patch is much less important than how much of the tire is—or isn't—slipping.

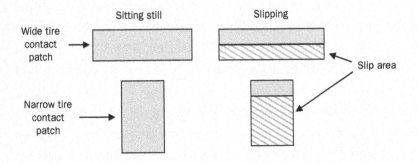

If we assume a thirty-six-square-inch contact patch, a tire at 10 psi would support only 360 pounds. In fact, if you add up the weight supported by all the tires at the minimum tire pressure for Martinsville, the total comes to only 2,340 pounds, which is much less than the roughly 3,600 pounds the car and driver weigh. You would need a contact patch of 55 square inches to support the weight of the car. Goodyear estimates that, at some tracks, the right front tire can have 3,400 pounds pushing down on it, which would require about 94 psi of pressure.

Some of the weight is supported by the sidewalls. Racing tires have much stronger sidewalls and lower aspect ratios than passenger-car tires. The aspect ratio is the tire's height (measured from the bead to the top of the tread) divided by the tire's width. Racing tires have aspect ratios around 52, while a typical passenger-car tire's aspect ratio is around 65. Low-aspect-ratio tires provide a stiffer and less comfortable ride, but comfort is not the object here—speed is.

Even with stronger sidewalls, having only 10 psi in the tires creates a lot of stress in the sidewalls. Swifty explained that tire pressures don't stay low for very long. Gases are made of molecules that are constantly moving. The air molecules around you move at an average speed of 1,100 mph. Tire pressure is created by gas molecules hitting the sides of the tire. Each time a gas molecule hits the inside of the tire, it exerts a tiny force. There are so many molecules in the tire that all those tiny forces add up to enough force to keep the tire inflated. The more gas you put in the tire, the more molecules there are, and the higher the pressure.

Once Swifty fills up the tires, the number of gas molecules doesn't change. When the car gets out on the track, the tires will heat up and this makes the gas molecules move faster. Faster-moving molecules exert more force when they hit the inside of the tire.

Put a blown-up balloon in your freezer and check back on it after about an hour. The balloon will have shrunk. The air pressure is less when the balloon is cold than it is when the balloon is warm because colder air molecules move more slowly. If you let the balloon warm back up to room temperature, it will return to its original size as the molecules move faster and exert more force on the inside of the balloon.

The rule of thumb on passenger cars is that tire pressure changes about 1 psi for every 10°F temperature change, which is why you should measure tire pressure when the tires are cold. Your tires reach temperatures of about 160°F on the expressway, but the tires on the No. 19 can reach temperatures of up to 300°F during a race. After two laps here at Atlanta, Swifty said, the right-side tire pressures will increase by about 11 to 12 psi and the left-side pressures by 7 to 8 psi. After a long run, the left-side pressures can increase by 10 psi and the right-side pressures by 20 psi. One of the most important things Swifty will do during the upcoming practice is measure tire pressures after the tires are removed from the car to find out how much of a "build" they experience. Starting at a lower tire pressure means that the pressure will be just right after the tire heats up.

Pressure changes the contact-patch size. An overinflated tire has a smaller contact patch. Blow up a balloon a little bit and set it on a table. The balloon is underinflated, so a lot of the balloon rests on the surface of the table. If you inflate the balloon more, it becomes rounder, and less of the balloon touches the table. A smaller contact patch means that the car's weight is concentrated in a smaller area, meaning less grip and less heat dissipation.

Although the larger contact patch produced by underinflating the tire creates more friction, an underinflated tire also heats due to flexing. Stretch a rubber band back and forth ten or twenty times, and then hold it to your lip. The rubber band will have gotten warmer. Tires do the same thing.

Deforming a tire takes energy. That energy is converted to heat and is thus not available for making the car move faster. Lower-aspect-ratio tires minimize deformation, as does keeping the tires properly inflated. Underinflated tires on a passenger car can lower gas mileage by 0.4 percent for every 1 psi drop in each of the four tires.

I've been saying "tire pressure" instead of "air pressure" because after recording the tire information, Swifty removes the valve stem from each tire and lets out all the air the Goodyear technicians put into them. He then refills each tire with ultra-high-purity dry nitrogen from a gas cylinder.

Air is 78 percent nitrogen, 21 percent oxygen, and 0.9 percent argon. The remaining 0.1 percent is a mixture of other gases. Air is mostly nitrogen already, so why bother with the 22 percent of air that's not nitrogen? The problem isn't the oxygen—it's water vapor that is often mixed in with the air. The pressure exerted by water molecules in the tire can change significantly near water's boiling temperature of 212°F (100°C). Small temperature changes in that region can produce large pressure changes. Races are sometimes won and lost by thousandths of seconds, so a team can't overlook any possible advantage.

All the above arguments tell us, however, is that whatever gas you use in your tires should be as free of water vapor as possible. So why not just use dry air? Dry air usually is produced by passing regular air through a compressor and a dryer, which removes some—but not all—of the moisture. You could purchase dry air; however, you can get dry nitrogen with a third less water than dry air for the same or slightly less cost. NASCAR doesn't mandate what gas to use in the tires, but you'd be hard pressed to find anyone using air on pit road.

That doesn't mean that everyone uses nitrogen, although the alternatives are slim. Only ten of the naturally occurring ninety-two elements in the periodic table are gases at room temperature. We can eliminate gases that are flammable (hydrogen), poisonous (chlorine), or radioactive (radon). The most promising of the remaining gases is argon. You can buy argon with about one-quarter of the water vapor in nitrogen, but it's about twice the price. Swifty told me that some teams use argon or an argon/nitrogen mix, but that most don't find enough of a performance difference to bother with the extra tanks.

Regardless of the type of gas in the tire, a blow out can be a race-ending or potentially more serious event. It's bad enough to lose a tire doing 65 mph on the highway. Imagine what it's like to have a flat at 175 mph with other cars just a few inches away. The first tires invented didn't go flat because they were solid; however, a trip over a cobblestone road on a solid tire was not very comfortable. The original tires for horse-drawn carriages were wood or wood wrapped with metal. Vulcanization made the solid rubber wheel possible, which had a little more give than the wood-and-metal wheel it re-placed.

In 1845 (prior to the invention of both the bicycle and the auto-mobile), the Scottish engineer Robert Thompson invented the pneu-matic (meaning air-filled) tire for horse-drawn carriages. His tire was an inflatable canvas tube surrounded by a leather outer tire. This tire produced a more comfortable ride, but was so difficult to manufacture and fit to the wheels that it was more trouble than it was worth. Thompson's invention of the fountain pen was much more appreciated.

In 1887, a complaint from his tricycle-riding son inspired veteri-narian John Dunlop to experiment. Dunlop learned that doctors gave patients air mattresses to make them more comfortable, so he put a rubber tube around a tricycle wheel and wrapped the entire assembly with linen tape. He then inflated the rubber tube with air and the pneumatic tire was (re-)invented.

The modern automobile wouldn't be possible without the pneu-matic tire. The ride would be so uncomfortable that people would walk. Even Kenny Schrader, who will race anything, would balk at 500 miles in a stock car with solid tires. The best compromise is a pneumatic tire that goes flat in a way that allows the driver to retain some control over the car.

Watching Swifty make last-minute adjustments as we got closer to practice time, I noticed that he had to unscrew two caps for each tire. Each tire is actually two independent tires: an outer tire, and an inner liner that is pressurized from 12 to 25 psi higher than the outer tire. If the outer tire loses pressure due to a puncture, the inner liner remains pressurized, almost always allowing the driver to get the car to the pits safely. While these tires still go "flat," they make a flat tire a manageable event instead of a race-ending catastrophe. NASCAR requires the inner liner at all tracks longer than a mile, as well as on the right-side tires at Bristol, due to the high banking that places a lot of stress on the right-side tires.

Josh and Swifty had decided which tires to start practice with and the tire pressures they would run. As Swifty sorted through his tires, I mentioned that he's moved more tires than I've had on all the cars I've ever owned—and it's not even noon.

"If I had a dollar for every tire I've handled . . . ," he laughed, hoisting the set of four tires on a cart. He wheeled the tires across the lane, around a fan taking pictures, and into the garage, where the No. 19 car was waiting in anticipation of prequalifying practice.

Eleven
Shock Therapy

"Watch," Chris, the hauler driver for the No. 19, said. "Drivers and management never close the hauler doors."

Open doors aren't usually a big deal, but we really wanted to keep the cold wind out. There are advantages to being a shock specialist on a frigid day like this, because all of the shock specialist's equipment resides inside the hauler.

Kevin "Kiwi" Duncan is the No. 19 team's shock specialist. As his nickname indicates, Kiwi is from New Zealand. Kiwi barely edges out Swifty as the senior member of the garage crew. Swifty looks like a tire specialist but Kiwi—with a little less hair and a little more gray—looks more like a mild-mannered dad (which he is). He used to travel around the world with the Champ Car series, but, he shrugged, "Now I go to Bristol and Martinsville."

Kiwi's title is shock specialist, but he also is responsible for the springs and sway bars that, along with shocks, are the most easily adjustable parts of the car's suspension. The suspension gets its name from passenger compartments on horse-drawn buggies that were "suspended" from the frame with a rubber band–like contraption. The suspension made the ride more comfortable by insulating the occupants from bumps and dips in the road. Comfort isn't an issue for race cars. The suspension is there solely to maximize grip.

The front and back suspensions are configured differently because the rear wheels drive the car while the front wheels steer. The front wheels are independent of each other. The spindle on which the wheel rotates is connected to a steering knuckle. Two control arms, one at the top and one at the bottom, connect the spindle to the frame.

The control arms are sometimes called "A-arms" because they're shaped like the letter *A*. The pointy tips of the A-arms are attached to the steering knuckles with ball joints that allow them to move left and right, up and down. The two legs of the control arms attach to the frame with hinges. The springs rest on the lower control arms and are held in place on top by a horizontal plate attached to the frame with a threaded bolt that can be tightened or loosened to change the spring's resting length.

The rear suspension uses trailing arms (also called truck arms), which are long pieces of metal I-beam that run from about the midpoint of the car, where they are connected to the frame with hinges, to the rear axle. The trailing arms fix the position of the rear axle front-to-back, but allow the axle to move up and down. The rear springs rest in pockets on top of the truck arm, just ahead of the axle, and, as in the front suspension, are held in place by adjustable plates attached to the frame.

Hold your arm straight out from your body and ask someone to grab your wrist and gently twist. That's torsion, which is how springs work. Torsion is a type of stress, like that caused by pushing or pulling, but torsion is a twisting type of stress. A long metal rod, for example, might be easy to bend but hard to twist.

A spring is a metal rod that's been coiled up. When you push down on a spring, the wire from which the spring is made twists and resists being pushed. NASCAR mandates the spring diameter, length, and number of coils, with different requirements for front and back springs. Kiwi opens a door in the hauler to reveal the team's spring collection. These are pretty big springs: The rear springs are around five inches in diameter and eleven to fourteen inches long, while the front springs are about a half-inch larger in diameter, but shorter in length by four to five inches. The wire from which the springs are made—which is more than a half-inch in diameter—is a material called spring steel, which is extra resilient. This resilience prevents the spring from developing too many defects as a result of the constant cycle of compression and release. Most spring steels are the standard carbon and iron, with chromium and either silicon or vanadium added.

Springs are characterized by their spring constant, or "rate," which tells you how much the spring resists being stretched or compressed. When Josh asks for a "600-pound spring," he means a spring with a rate of 600 pounds per inch. A 600-pound-per-inch linear spring will compress one inch when 600 pounds are applied, two inches with 1,200 pounds applied, and so on. The springs in the hauler range from 300 to 3,000 pounds per inch.

Progressive springs have unevenly spaced coils or coils with different diameters. If it takes 400 pounds to compress a progressive spring by one inch, it takes more than 800 pounds to compress it two inches. NASCAR doesn't allow progressive springs because it increases costs for the teams and inspection hassles for the NASCAR officials.

Although he can't use progressive springs, Kiwi can change the way the spring compresses by inserting one or more spring rubbers. Spring rubbers are pieces of hard rubber placed between the coils that increase the spring rate by making it harder to compress the spring. It takes too long to change a spring during a race, but a spring rubber can be added or removed quickly.

Spring rubbers come in soft, medium, and hard, but Kiwi suggests they ought to be called "hard, harder, and really hard." Harder rubbers increase the spring rate more. Kiwi has full, half, third, and quarter rubbers. A full rubber makes the spring more progressive than a fraction of a rubber. The car can't start with spring rubbers in the front springs, but teams can start with rubbers in the rear springs for added adjustability during the race. Some spring rubbers have handles to help the pit crew remove them quickly.

The spring compresses when the wheel goes over a bump. A weak spring provides a smooth ride because the wheels move over the bumps and the rest of the car stays relatively still; however, the car doesn't respond as quickly because the frame isn't connected tightly to the wheels. Stiffer springs make a car more responsive, but the ride is bumpier.

The spring on a car's wheel behaves much like the spring on a bobblehead. If you push down on, say, a Tony Stewart bobblehead, you convert motion energy into stored spring energy the same way a spring in a car does when compressed. When you let go of Tony's head, it bounces up and down. The energy is converted from stored spring energy to motion energy and back again, until all of the energy in the spring is dissipated by friction. While this up-and-down motion is the whole point of a bobblehead, you don't want to see your car doing it on the track.

When a wheel hits a bump, the spring absorbs the energy by compressing, but you need to transfer that energy elsewhere to prevent bouncing. If you push the back of your car down and let go, the bouncing subsides pretty quickly. The shock absorber, which is also called a damper because it reduces or dampens the oscillatory motion of the spring, converts the stored spring energy into motion energy (the movement of the shock's shaft), and finally to heat energy that is dissipated into the surrounding air. Shocks are mounted between the control arms and the frame in the front and between the trailing arms and the frame in the back.

Shocks and springs respond differently, which Kiwi demonstrated using the shock dynamometer in the hauler. A shock dyno (which also is used for springs) is about three feet tall with two one-inch-diameter steel rods rising vertically from its base. A crossmember runs across the two vertical rods and a hydraulic activator rises from the center of the base. Kiwi placed a purple spring (springs are coated to prevent rusting) on the base and locked down the crossmember to hold it firmly in place.

The dyno compressed the spring and a load cell measured the applied force and how far the spring compressed. A computer generated a plot of force versus distance, which was a straight line (as it should be for a linear spring). If the force doubles, the amount by which the spring is compressed doubles as well.

Unlike a spring, a shock's resistance isn't proportional to how far you pull or push: It's proportional to how *fast* you pull or push. The faster you move the shaft, the more it resists moving. A shock is a hollow cylinder. A floating piston (an aluminum disc that forms an O-ring seal with the inside of the shock) at the top separates nitrogen gas from oil. One end of the shaft sticks outside the shock and the other end extends into the oil. The oil resists pulling or pushing the shaft.

Kiwi fastened a shock on the dyno. The dynamometer pumped

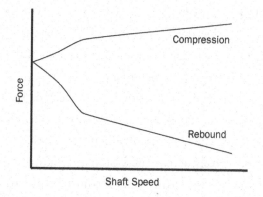

the shaft up and down to measure the resistance at each speed. The plot of force versus speed that the computer generated for the shock Kiwi was testing appears in the graph on the previous page.

When the tire hits a bump, the shaft is pushed into the shock, which is called compression. Rebound is the shaft being pulled out of the shock. Rebound controls how fast the car returns to its normal height after the spring is compressed. Lots of rebound holds the car to the track for a longer time after the spring compresses.

The specific behavior is determined by what's inside the shock. Josh asked Kiwi to change the rebound, which gave Kiwi an opportunity to show me how to build a shock. He clamped the shock—which is almost two feet long when fully extended—into a jig on his workbench and fit a plastic collar around the lower part of the cylinder. After letting out the gas that pressurizes the oil, he used a wrench to unscrew the two halves of the shock. The collar caught the oil that overflowed when he extracted the shaft. Unscrewing a nut on the end of the shaft released the piston and two groups of shims, which are thin metal disks that look like washers.

Shim thicknesses range from about two thousandths to twenty thousandths of an inch. A typical washer from the hardware store is a sixteenth of an inch (around sixty thousandths of an inch) thick. Shim diameters run from a little less than an inch to about 1³/₈". A stack of shims is placed on each side of the piston. The rebound stack sits on the top and the compression stack is on the bottom, as shown in the drawing on the next page.

When the shock shaft moves slowly (one inch per second or less), the shims don't really do anything. The piston has a complex arrangement of holes, the majority of which stay covered by the shims. Most of the oil travels across the piston through small holes that aren't covered by shims, or through an adjustable bleed valve in the

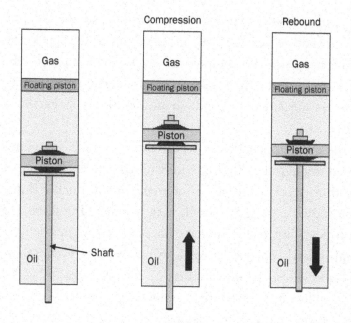

piston shaft. A cone-shaped protrusion called a needle valve can be adjusted up or down to control how much oil flows through the bleed valve. The needle valve is opened or closed by an adjuster at the bottom of the shock that clicks when turned.

"That's the only thing we can really change during a race," Kiwi explained as he wiped the oil off the shims he had just removed. "In some racing series, you can have four or five different adjustments, but NASCAR only allows one adjustment per shock." The amount of change is denoted by the number of "clicks." Crew chiefs normally don't adjust the shocks during a race unless the car is handling really badly.

If the piston moves fast enough, the shims bend, opening paths for oil to flow through the piston. Faster movement makes the shims bend more and opens more paths for the oil to flow through. When the shaft moves in compression, the upward motion holds the rebound stack flat against the piston and the compression stack bends. When the shaft is pulled out of the shock, the opposite happens: The compression stack closes and the rebound stack bends open. Thicker shims require higher pressure and thus faster piston speed to bend, so thicker shims make the shock stiffer.

The shim stacks allow Kiwi to tailor the shock's response. He sorted through the shims he had removed and changed out one of the thin pieces of metal for another, slightly thicker shim, and then started putting the shock back together. The shims provide great flexibility. Kiwi can make linear stacks, where the force depends linearly on speed; progressive stacks, where the shock gets stiffer as the shaft moves faster; or digressive stacks, which are stiffer at low shaft speeds than at higher speeds. These characteristics can be combined: for example, linear compression and digressive rebound.

Kiwi carefully recorded in a spreadsheet the shims he used. After tightening the shims on the shaft, he refilled the shock with oil and reassembled the two pieces. The final step was pressurizing the shock with nitrogen gas, which prevents the up-and-down motion of the piston from creating bubbles in the oil. Air pockets make the oil respond differently, because the bubbles have to be compressed before the oil can offer any resistance.

Shocks can reach temperatures of 180°F to 200°F due to a combination of friction between the piston and the oil and radiant heat from the brakes. The oil, which usually is 2- to 5-weight, becomes less viscous as it heats. Decreasing viscosity leads to "shock fade," a condition in which the shock gets less stiff as the oil's viscosity decreases.

Kiwi tested the shock on the dyno and overlaid the results on the previous graph to see if it met Josh's specifications. It looked pretty close to me, but Kiwi wondered aloud if it was close enough. Wondering anything aloud in a hauler is a dangerous proposition. Swifty, who had ducked in to warm up, started giving Kiwi a hard time about it not being *exactly* right, which prompted Kiwi to take apart the shock again. He changed another shim, which made the resulting dyno curve acceptable, even to Swifty.

Shocks and springs were originally designed to improve mechanical grip—how much the car's tire is pushed into the track—but their roles changed as aerodynamics became increasingly important. The springs set the car's ride height. When the car accelerates, brakes, or corners, the weight shifts and compresses the springs, which changes the car's height above the ground and its attitude.

You can't have the bottom of the car hitting the ground when you corner, so the traditional setup used a very stiff right-front spring—1,800 pounds per inch was not unusual—to hold the car off the ground going into turns. Crew chiefs realized from wind-tunnel experiments that springs could be used to keep the car in an aerodynamically advantageous position. Announcers talk about a "hound-dog attitude": the car's nose is down and its tail is up. One way to get the car in this position is to use soft springs in front; however, you have to keep the car from bottoming out.

If you use very soft front springs—say, 200 pounds per inch—the front springs will compress so much that the coils actually touch each other, which is called binding. A coil-binding setup makes the spring rate effectively infinite when the coils touch. Stiffer rear springs keep the tail of the car in the air to generate downforce on the rear. Some mechanical grip is sacrificed for the aero-grip that comes from having the car in the most aerodynamically favorable position.

Although this setup sounds simple, getting it just right isn't. The driver goes from riding normally to effectively not having a spring at all, which can lead to unpredictable behavior. How (and how quickly) the coils touch is critical. Coil binding creates more variations in the load on the contact patch, and a tire with a coil-bound spring has to absorb all of the energy from bumps—which tires really weren't designed to do.

The new car, with the splitter sitting so close to the bottom of the car, presents different setup challenges. The new car has less travel, meaning that the difference between the car's maximum height and the car's minimum height is smaller. Coil binding with the new car is more difficult.

NASCAR allows teams to use bump stops on the new car, which are hard pieces of rubber that fit either around the shock shaft (where they are sometimes called "packers"), or between the frame and the A-arm (where they are called "chassis stops"). Bump stops serve the same purpose as a bound coil: They keep the car from bottoming out. Instead of riding on the coil-bound spring, the car rides on the bump stop.

Satisfied with the shocks, Kiwi looked at his watch. Josh had asked Kiwi to be ready to make a sway-bar change during practice. Sway bars (also called torsion bars, anti-sway bars, roll bars, or anti-roll bars) make the two sides of the independent front suspension a little less independent, which constrains the weight roll through the corners.

A sway bar is a steel rod about three feet long. Each end is connected to an arm that is about a foot long and runs perpendicular to the bar. The ends of these arms attach to the lower control arms on each side of the car, with the bar itself passing through a tube in the frame parallel to the front bumper.

If both tires move upward at the same time, the sway bar doesn't do anything: There has to be a difference in motion on the left and right sides. When the car goes into a left turn, the right-front wheel moves upward and the right-front spring compresses. Without a sway bar, the compression would encourage the body to roll toward the outside of the turn. The sway bar links the motion of the two wheels. The upward motion of the right wheel twists the sway bar. The twisting sway bar raises the left wheel and compresses the left spring, which keeps the body more level. If the sway bar is too stiff, going over a bump with one tire can lift the other tire off the track.

You could decrease roll using very stiff springs, but stiff springs don't keep the tires on the track as well. Resistance to twisting makes a sway bar stiff. Sway bars range from an inch to two and a half inches in diameter. Some are solid and some have a hole running the length of the bar. Solid sway bars are stiffer than sway bars with holes, and larger-diameter sway bars are stiffer than smaller-diameter sway bars. In most cases, a stiffer sway bar makes the car tighter.

Kiwi is meticulous. Everything in his station is labeled and in its appropriate compartment. Spilled oil is cleaned up quickly. He's explained things well enough that I could probably fill in for him in a pinch. He grabbed the new shock and pulled on a coat and a stocking cap.

"The only time I had trouble with a shock during a race," he said, "is the one time I didn't install it myself. So I like to do it myself. At least then I have only myself to be mad at if something goes wrong."

Kiwi headed over to the garage to install the rebuilt shock. He carefully shut the hauler doors behind him.

Twelve

The Two-Lap Dash:
A Qualified Success

Josh and Chad (the team director and engineer) have been sequestered in the back of the hauler from the moment the team arrived at the track this morning. I peeked into their command center, which is a six-by-ten-foot space at very back of the hauler. The only decorations are driver Elliott Sadler's firesuit hanging outside a full-length cabinet and a plasma TV screen on the wall to my left. A long countertop that runs along the same wall and curves to form an "L" with the back wall serves as a desk. A padded vinyl bench that can seat four fills the length of the right-side wall.

Josh and Chad were reviewing notes from previous Atlanta races on their laptops. There were a lot of decisions to be made between now and practice—the only practice before qualifying—at 3:30 that afternoon. The Weather Channel was on and the line of storms moving out of the area was especially welcome news. If qualifying was rained out, cars would start in order of owner's points, which would put the No. 19 back in 34th position.

The area just outside the hauler serves as a sort of porch from which you can watch the NASCAR world go by and be mostly out of the way. I parked myself there when the inside of the hauler got busy. This morning, a woman and her daughter stopped by. The daughter, a pretty blond in her early twenties, met Elliott at the track last year and hoped to see him again. During our conversation, her mother made a comment I would hear a couple of times this weekend.

"*I don't understand why they keep giving Elliott the crap cars.*"

Josh rolled his eyes when I asked about this comment later. His answer is a mix of frustration and irritation.

"Look, it's not that simple. People just don't get it," he said, shaking his head.

The television analyst Larry McReynolds—a former crew chief— likes to refer to race cars as "science experiments." I think it's an especially apt analogy, except that most people have never done a *real* science experiment. The things called labs in school give you a set of directions. You follow the instructions and write a conclusion that you knew before you even started.

That isn't an experiment. An experiment is asking a question you *don't* already know the answer to. The question for today—for every one of the races this season—is, "How do we make the car go faster?"

Josh was still a little agitated about my question. "Yes," he said, frowning, "there are inevitable differences between equipment, but they are small differences."

"Your car can have a few-horsepower advantage," Josh shrugged, "but it will do you absolutely no good without the right setup."

The setup is the things on the car you can change at the track, including springs, shocks, sway bars, tire pressures, and weight distribution. With so many things to change, setting up a car should be easy, but the number of variables is actually what makes it so difficult.

There are four springs on the car, one on each wheel. Assume that Josh can choose from two different springs (which I'll call A and B) for each wheel. If Josh picks A for the left front, he has eight possible unique configurations, as shown in the table below. If he picks B for the left front, he has eight more configurations, giving him a total of sixteen different configurations. If Josh has four spring choices for each wheel, there are 256 possible combinations. Looking at the collection of springs the No. 19 team has brought with them, I estimated that Josh had—at minimum—six spring options per wheel, which gave him almost 1,900 different possible combinations—*and that's just the springs.*

Left Front	Right Front	Left Rear	Right Rear
A	A	A	A
			B
		B	A
			B
	B	A	A
			B
		B	A
			B
B	A	A	A
			B
		B	A
			B
	B	A	A
			B
		B	A
			B

Even this explanation oversimplifies things because some variables are correlated to each other. Changing one affects the others. Changing the springs, for example, changes the weight distribution and may require different shocks.

The labs you did in school probably taught you that the proper way to conduct an experiment is to hold all the variables (the things you can change) constant—except for one. Changing the value of that one variable and repeating your experiment lets you figure out its optimal value. This is a great idea in principle, but it doesn't work for a race car—or any other complex system.

Let's go back to having just two choices for each spring, which gives you sixteen configurations. Working their fastest, Josh estimated it would take them more than two hours to try all the combinations, and again, that's only the springs. The upcoming practice would only be an hour and a half long.

Luckily, Josh and Chad have tools to help them narrow down the number of variables and the possible values for those variables. Crew chiefs have always relied on extensive notes from previous races and test sessions that document how the car (and the driver) respond to different setups. Unfortunately, the only other race the team had run with Elliott at Atlanta was the worst finish of all the races they've run together.

"Well," Chad said, "at least we have some idea what *not* to do."

Josh noted that last fall's qualifying at Atlanta had been rained out, but that Kasey had won the pole for the spring race.

"And," he mused to Chad, "his second lap was faster than his first."

Chad looked over at the page of scrawled notes scanned into the computer and acknowledged the observation with a quiet, "Hmmm."

Another tool Josh and Chad have is sharing information with the other Gillett Evernham Motorsports teams. Chad explained that he and Josh are connected to their counterparts in the No. 9 and No. 10 haulers via a computer network. Each team can access the other teams' setups and they can discuss things in a real-time private chat room.

While Josh typed notes to the No. 10 team, Chad was working with one of the more powerful tools in their arsenal. He can enter two or three sets of values into a program and the program will predict for him how the car should respond. Refreshing the screen generates graphs that compare the effects of different sets of variables on quantities like turning force.

This program utilizes a principle developed for systems with large numbers of correlated variables. There are simply too many possible combinations of variables to test each one. You need a way to determine which variables are most important under different circumstances, how those variables depend on each other, and the best values for those variables. The experimental procedure is called "Design of Experiments," or DoE.

Josh explained that there are about twenty-five relevant variables for each track; however, different variables are important at different tracks. Like most intermediate tracks, aerodynamics is very important at Atlanta, but the rough surface makes tire wear rate important, too.

Even with all these tools, finding the right values for all the variables is like navigating a complex terrain using a map drawn by someone who has never been there. Imagine being dropped into a landscape with a series of heavily forested rolling hills and valleys without a map. I've drawn one possible landscape on the next page to illustrate. The "perfect" setup corresponds to being in the lowest valley (point B). You have to figure out how to get there.

Let's say you land at point A, which is a valley, but not the deepest valley. You look to your left and your right, and you realize that it's uphill in both directions.

If you had an accurate map, you'd know that you're not in the lowest valley and that you have to go uphill to get to that ideal setup. But none of the teams have a map—only a driver telling them if he thinks they are going uphill or downhill.

A driver happy with his car from the time they unloaded probably landed in a valley like point B. The driver who complains that nothing they do to the car makes it better is probably at a point like C, where you can go a long way and not be able to tell whether you're getting any closer to a valley.

It would be fair for Josh to interrupt now and reiterate his point, "it's not that simple." He would be right. My picture corresponds to a single variable—like one spring or the pressure in one tire. Adding another variable would make a three-dimensional landscape. What do you do when you land at the center of a Pringles potato chip–shaped feature? If you travel in one direction, you go up, but if you travel perpendicular to that direction, you go down. Inside the hauler, Josh and Chad were trying to navigate around a space with twenty-five dimensions, which is impossible to draw—and even harder to imagine.

Just in case you're not convinced of the complexity of setting up a car yet, there are variables over which the team has absolutely no control. Practice would be at 3:30 that afternoon, but qualifying wouldn't start until seven. The temperature, the humidity in the air, and the amount of rubber on the track will be different. The landscape they are trying to navigate will change between practice and qualifying. Peaks and valleys may shift up or down, and the perfect setup from the end of practice may not be the ideal setup—or even a good setup—for qualifying.

Let's say that, at the end of practice, the setup corresponded to being in valley B in the figure below, which is the best they could do—at that time. When qualifying rolls around, the landscape has changed and position B is no longer the best setup. A team that ended practice in setup C might be happier: When the track changed, that giant plateau on which they were stuck shrank. All of a sudden, the car is handling better than it was before, even though they didn't change anything.

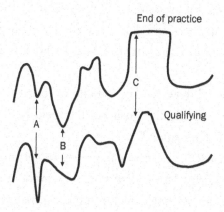

The quest for the perfect qualifying setup continued back in the command center. Each car will run a maximum of two laps, with the fastest lap determining where that car starts the race. Theoretically, only the fastest forty-three cars get into the race, but the top thirty-five cars in owner's points are guaranteed a starting spot regardless of qualifying time. Past NASCAR Sprint Cup Series champions have a provision allowing them to race six times per season, even if their qualifying times aren't fast enough. If there is more than one champion, the most recent has dibs on the provision. Even though the team is guaranteed a spot in the race, they want to start as close to the front as possible.

The No. 19 car is unloaded in race trim, which is their initial guess at the setup they will use during the race. If things go well, they will make minor tweaks to the setup and then change to qualifying trim for a few runs. The "go or go home" cars—those that have to earn their way into the race by posting a fast-enough qualifying time—unload in qualifying trim. If they don't qualify well, they don't race.

At about 2:45 P.M., Josh shouted—to no one in particular—"Anyone seen The Driver?" Not referring to The Driver by The Driver's name seems to be a tradition. Chad started to note that it was still early, but he was interrupted by Kirk Almquist, the car director, who bounded up the stairs. Kirk is a twenty-eight-year-old native Californian and looks the part: tall, blond, and shivering in a company parka.

"Best news you'll have all day," he said to Josh.

"We're going out fiftieth?" Josh asked hopefully.

"Forty-sixth," Kirk smiled.

The order in which the fifty cars trying to get into the race will make their qualifying laps is determined by random draw. Today, a later draw is better. A cooler track has more grip. Qualifying late is always an advantage for multicar teams. The No. 10 will take its qualifying run 15th and the No. 9 car will go out 25th, which means that the No. 19 team will have additional data before they have to qualify.

Questions and answers got shorter and more to the point as the afternoon progressed. Josh and Chad continued to examine the effects of different variables in their simulation programs. At about three o'clock, one of the more important variables walked into the hauler wearing jeans and a hooded black Dodge Motorsports sweatshirt.

Six-foot-two, dark-haired Elliott Sadler is an avid outdoorsman who has more than thirty hunting dogs he says he can tell apart by their barks. Although the picture on the side of the hauler shows him clean shaven, he more often sports a goatee and mustache. He is a very popular driver, in part due to his thick southern Virginia accent, charming smile, and quick sense of humor. When he learned I was writing a book, he asked whether it has to be correct. Assured that it does, he flashed a smile, grinned at the crew, and asked, "And you're talkin' to Josh?"

Television gives you the impression that a race team is the crew chief and driver, with a few people to change tires and add wedge, but there's much more to it. While the crew is in the garage preparing the car, Elliott is fulfilling a plethora of other responsibilities: personal appearances, signing autographs, meeting with sponsors, and doing interviews.

I don't mean to suggest even remotely that the only qualities required of a driver are good looks and a glib tongue. Elliott plays a pivotal role that goes well beyond what you see on television. Driving skill is important, but so are leadership and the ability to communicate with the crew. NASCAR doesn't allow teams to collect data directly from the car during race weekends, so Josh and Chad can't measure whether the turning force on the tire changes the way their simulations predicted it should. They rely on Elliott for that information.

The importance of communication can't be overemphasized. Josh and Chad, because they both come from formal engineering backgrounds, already have a common vocabulary. Josh and Elliott are still learning about each other. Although they both have extensive racing experience, the overlap in their backgrounds is much smaller.

They are working hard to bridge that gap. Chad told me that Elliott "is one of the smartest drivers in the garage." Josh added that Elliott is interested in learning about what engineering can and can't do—not all drivers are.

As practice time neared, we vacated the control center so Elliott could change into his firesuit. Standing atop the hauler is a good place to watch practice and be out of the way, so I scrambled up the ladder after Josh and Chad. Chris had set up a monitor connected to NASCAR scoring that shows all the cars' lap times. A metal shield around the monitor would have kept the sun's glare off the screen—if there had been any sun. It was still overcast, cloudy, and cold.

It was even colder at the top of the grandstands where Elliott's spotter, Brett Griffin, was standing. Brett helps Elliott on the track by telling him where other cars are and when it's safe to pit. He also talks to other drivers' spotters if a deal needs to be brokered or a message delivered. A spotter has to provide detailed information in a very concise manner. Brett demonstrated this skill with a colorful description of exactly how cold it was on the grandstand as a check of the radio connection.

At 3:30, Brett announced, "The track is hot," and the cars roared out. The crew gathered around the monitor on the toolbox in the garage to watch the lap times.

Elliott was pretty happy with the car the first time out and, after a few laps, headed for the garage to try some changes. Josh scrambled down the hauler ladder, followed by Chad, who carried a tablet notebook computer into which he entered all the changes and Elliott's comments about them.

Elliott made two more runs in race trim and then Josh called for a switch to qualifying trim. Thirty-five minutes into practice, there were more cars in the garage than on the track. Most of the cars on the track were those that had to qualify on time and were already in qualifying trim.

Qualifying trim is optimized for speed. Additional tape on the grille restricts airflow to the engine, but provides more front downforce and more grip. You can't run a whole race with the grille taped up, but it will be fine for two laps. The change in front downforce requires changes to the rest of the suspension, so shocks and springs are switched out. The mechanics cover the air-cooling ducts to make the car even more streamlined and bypass the oil cooler to reduce the load on the engine and give the car a little extra horsepower.

Chris poured bags of ice into a cool-down machine that pumps cold water through the engine. The car will run two laps at maximum speed. If the engine starts out hot, it's just going to get hotter, especially with some of the cooling mechanisms bypassed. A cooler engine also has more horsepower because cooler air is denser, so more oxygen molecules are pulled into the cylinder. It's not a huge gain—maybe 3 to 4 horsepower—but depending on the track, that could make the difference between qualifying on the pole or back in twentieth place.

Once the changes were completed, Elliott—who had been watching other drivers' lap times intently on the computer monitor—went out to make a practice qualifying run. His first lap indicated a definite need for changes. Josh and Chad didn't even wait for him to finish the second lap before they started climbing down from the hauler. These changes wouldn't take long, so Elliott waited inside the car with his helmet on.

"Anyone know how fast I am and how fast I need to run?" Elliott asked over the radio, sounding a little irritated that no one had told him yet.

Josh chuckled. "I know how fast you ran and how fast you need to run. You ran a 58 and the fastest time was a 35. We're at P13 now." A "58" is a time of 30.58 seconds. P13 means that they had the thirteenth fastest lap.

The changes were completed about 4:40 and Elliott zipped out of the garage. He circled around to start his mock qualifying run, but by turn 3, he was already slowing down and heading for pit road. I looked over at Josh to try to figure out what was happening. Josh grimaced. Chad looked down at his feet. Brett's voice crackled over the radio.

"Almost wrecked."

As I said, spotters are trained to be concise.

Elliott was also concise. He exhaled loudly as he headed for the garage. Josh keyed his mic. His grimace had turned into a frown, but you wouldn't have known it from his friendly tone. "What, Bud?" he asked Elliott. "I didn't catch that."

"Let me catch my breath," Elliott said tersely, "and then I'll talk to you." I would be terse, too, if my car almost went out from under me at 180 mph.

Josh, who was still on top of the hauler, suggested making a change and trying another run, but Elliott quickly vetoed that idea, saying that there wasn't time to make changes and get out again.

The No. 19 garage was not a happy place. Josh dropped his clipboard loudly, climbed down the ladder and disappeared into the hauler, followed by Chad, still trying to enter information on the computer. Elliott emerged from the garage and went directly to the hauler, shoulders hunched over, eyes down and clearly not very happy. When practice ended, they were at P20, with a best lap of 29.582 seconds. Kurt Busch posted the best lap at 29.165 seconds.

A few moments later, the blond and her mother came by and were disappointed to find that Elliott had already finished practice. They waited in front of the hauler. About twenty minutes later, Elliott—back in jeans and sweatshirt and looking no happier than he looked when he went in—emerged, still tight-lipped and eyes cast downward. He hurried out of the trailer right past the young woman. She called out to ask for a picture, and by the time he rotated around so that she could see his face, he was wearing a captivating smile. Elliott put his arm around her and made small talk while her mother took a picture. The young woman was ecstatic and thanked him, wishing him good luck in qualifying. As soon as the visitors turned to leave, Elliott stuck his hands back in his sweatshirt pockets and, head down, walked quickly off to his motor coach.

I cautiously ventured back into the hauler to find Josh and Chad focused again on their laptops: Even if they didn't get what they most wanted from practice—P1 on the speed chart—they got the next best thing, which is new data.

Josh was happy to get what he calls "objective data." Lap times and tire wear measurements aren't subject to interpretation. Chad started typing information into his computer while Josh went to meet with his counterparts from the No. 9 and No. 10 cars to debrief. The pace in the hauler and in the garage quickened.

Chad conferred briefly with Kirk and the garage started buzzing again. The car had to be in line for prequalifying technical inspection by 6:20 P.M. Since we would go out late, we didn't mind being at the end of the line. At 6:30, Josh and Chad were still in the hauler, running different scenarios on their computers. Josh turned on the television just after 7:10 to watch the first cars qualify.

"Riggs is in P2," Chad said of the No. 10 car's driver. Chad lent me a parka—it was about 35°F at this point—and we headed to pit road. The rest of the team was already there with the car. Chad and Josh talked to Rodney Childers, the No. 10's team director, and then Josh radioed Swifty to stand by in case he wanted a right-side tire-pressure change.

"Elliott called me this afternoon and suggested we run two laps," Josh told the team over the radio with a trace of amusement.

Mark Martin had the best lap so far, with a time of 28.909 seconds. The No. 9 car turned in a lap of 29.074 seconds, which put them in fourth position in the ever-changing lineup.

The car continued to inch along pit road. We followed and watched Jeff Hammond interview Elliott. Josh was nervous, cold, or both. He paced, jumped up and down a few times, and kidded with Elliott the way you do just before you have to give a presentation or take a big test. When we were the fourth or fifth car from the front of the line, Elliott got into the car and donned his safety gear. Ramon "Razor" Zambrano, another of the mechanics, cleaned the windshield one last time. Josh motioned Swifty over and ordered a last-minute tire-pressure change—he didn't want to say how much over the radio. Swifty made a final circuit around the car, checking the pressure in each tire and letting out a little air from one right-side tire.

I sat on the edge of the pit wall with the team as Elliott went out for his first lap. I knew that everyone was holding their breath because it was so cold that you could see when people breathed. There were no clouds of condensed air anywhere around our group.

Elliott roared out on track. His first lap was 29.631 seconds, which put him in a disappointing 42nd place. Josh was surprisingly unconcerned. Elliott completed his second lap and we all cheered: 28.891 seconds, which, Josh told him happily, was good enough for P2 at an average speed of 191.894 mph.

"I'm sorry, guys," Elliott said as he pulled back on pit road. "I left a little on the track." Josh shook his head. "That car didn't have another two tenths in it," he said, which is what we would have needed to unseat Ryan Newman from the pole.

The top three cars are held on pit road to go through an abbreviated tech inspection one more time after qualifying ends, so we waited while the last five cars took their qualifying laps. None of the remaining cars beat Elliott's time. We would start in second position, ahead of Jimmie Johnson by just sixteen one-hundredths of a second.

Somewhere during today's events, I realized that I had switched from writing "they" to writing "we" when referring to the team. I tracked it back to just after Elliott almost crashed during practice. Even from the top of the hauler, I could see the crew members' shoulders sink and their heads hang. It made me happy to see them smiling and high-fiving now. So much for my intended impartiality.

It was nine o'clock. The team had been at the track for eleven hours, but NASCAR had just announced that the garage would stay open an extra half hour because qualifying had ended so late. Elliott explained the two-lap strategy that surprised only me.

"I got the idea from Robby Gordon—a race he ran a couple years back," he said, smiling. "The first lap is just to get the tires warm. Don't go for time, just get the car ready for the second lap. Everyone else drove hard the first lap and got the car mad."

Josh took a short phone call from his wife, and Sammy Johns, GEM's vice president of competition, stopped by with official congratulations.

This had been a successful science experiment. A lot of factors came together at the same time: the late draw, the cooler weather, Elliott's decision to run two laps instead of one, Josh's last-minute tire-pressure change, and, no doubt, a little bit of luck.

Ben Franklin noted, "I am a strong believer in luck, and I find the harder I work the more I have of it." Tonight, the No. 19 team made something that isn't simple look like it is. Josh leaned back in his chair and stole a brief moment to smile.

"And the world is good again," he said. Then he turned back to his computer in preparation for tomorrow. The garage opens at 8:00 A.M.

Thirteen

Keeping Drivers (and Fans) Alive

The NASCAR Research and Development Center is just across the street from Roush Fenway Racing in Concord, North Carolina. A race car occupies about half the small lobby. A man and his three boys wandered in and the boys—all teenagers or close to it—immediately headed for the car. The father went to the receptionist's desk.

"Is this the NASCAR Technical Institute?"

"No, sir," she said patiently, "This is the NASCAR Research and Development Center."

"What do you do here?" he asked.

"We work to ensure the safety of the sport," she told him, adding that the facility isn't open to the public.

"Ensuring the safety of the sport" doesn't begin to describe everything that happens here. The NASCAR R&D Center has three focus areas: making racing safer, keeping the sport competitive, and containing the costs of competing. NASCAR was the first racing sanctioning body to have its own R&D center. The 61,000-square-foot center opened in January 2003 under the leadership of Gary Nelson. Michael Fisher, a former auto executive from Michigan, became its managing director in mid-2006.

Mike and technical director Steve Peterson gave me a tour of the facility. Steve has more than twenty-five years of experience in the automotive-engineering industry, but his most significant contributions to motorsports are in the area of safety. Steve was awarded the Society of Automotive Engineers Motorsports Achievement Award in 2006, which recognizes his sustained record of leadership and contribution to motorsports.

The pristine white walls of the main rooms are two stories high, with smaller work areas defined by lower white cinderblock walls. The NASCAR R&D Center has similarities to race shops—for example, a small shop where they can build a complete car, as well as their own dynamometer.

Some race shops have museums where they display their accomplishments. The NASCAR R&D Center has its own museum of sorts, which is the caged second-floor area that lines one long side of the main work room. Parts, pit equipment, and even a couple of complete cars are visible through the chain-link front. This, I assumed, was the storied area to which innovations that fell in the wrong part of the "gray areas" of the NASCAR rule book were banished.

Among the items in lockup were three pit-road gas cans. I asked Mike, who had joined the R&D Center a little more than a month ago, how one could cheat with a gas can.

"I don't know," he said, turning to Steve. "How *can* you cheat with a gas can?" Steve—a no-nonsense engineer—got a pained look on his face.

"The rule," Steve said gruffly, "is that if it leaves the pit box, it's mine." It sounds like an imperious dictum someone made just because they could, but he went on to explain. Equipment usually leaves a pit box when the car drags it out. Although there might not be any obvious damage, safety may have been compromised. A gas can, for example, may develop a crack too small to see, but large enough to let flammable gasoline vapor escape. A damaged jack might let the car down before people are clear of it. Impounding the equipment ensures that it won't get used again and allows NAS-CAR to study how the equipment responds to damage.

Fans have come to expect drivers to walk away from spectacular crashes. After pirouetting twice before sliding across the infield and landing upside down at Daytona in 2003, Ryan Newman's major complaint was dirt in his teeth; however, we aren't that far removed from the days in which drivers didn't joke about crashes.

NASCAR has always made rules to ensure the safety of the competitors, but before the R&D Center, much of the work was done through the race teams, the manufacturers, and in the garages and basements of Gary Nelson and Steve Peterson. Manufacturers shared their extensive automobile safety research, but racing presents a unique situation because the energy scales involved are much higher. A race car going 180 mph has sixteen times the motion energy of the same car going 45 mph. If you used the motion energy of a race car at 180 mph to shoot a 150-pound person from a cannon, that person would travel almost five miles straight up.

Racing accidents do share one very important characteristic with passenger-car accidents: You can make the car as safe as possible, but it won't do you any good if the driver doesn't stop when the car stops. In 2007, NASCAR started requiring drivers to use a six-point restraint system. Some drivers were already using the six-point harness, while others had been using a five-point model. Both have two shoulder straps, each of which runs vertically over the driver's shoulders, and two lap belts that restrain the pelvis. The five-point belt has a single crotch strap coming up between the driver's legs that keeps the driver from sliding under the other belts, which is called submarining. The six-point harness has two belts that wrap around the legs, which is more effective than a crotch strap. All six belts come together in a quick-release lock near the center of the driver's pelvic region so that the driver (or rescue team) can release all the belts with one motion.

If keeping the driver in the seat were the only purpose of the harness, using padded metal instead of webbing might be more effective; however, the flexibility of the webbing has a purpose. When the car stops suddenly, the harness belts stretch a little, which extends the time over which the driver slows down.

The belts in race cars are woven much more tightly than those used in passenger cars so they don't stretch too much. Harness belts used to be made from nylon (the same type of material used for air bags), but most manufacturers are moving to polyester, which is as strong as nylon, slightly less stretchable, and doesn't degrade as quickly when exposed to UV radiation and heat. Environmental degradation is why restraint systems have expiration dates.

The harnesses work with the seat to keep the driver in the car, and they also guide how forces are distributed across his body. Drivers are fitted for seats like businessmen are fitted for suits: Both need to be tailored to the shape of the person using them. The seats literally envelop the driver, coming up high on the sides to surround and protect the entire pelvis. Broad arms wrap around the driver's ribs and a second set wraps around the shoulders. Two more arms extend on either side of the driver's head with about ¾" between the helmet and the padded arms. The harnesses prevent forward motion and the seat limits side-to-side motion.

The seat can't be too rigid because it needs to absorb some of the energy from the crash and not simply transfer it to the driver. The seat also can't be too weak because a seat damaged in an impact endangers the driver if there is a second impact.

Even without being involved in a crash, seats have to be strong enough to stand up to repeated cycles of vibration and temperature change. You can break a material with one large force, but also by repeatedly applying smaller forces. Fatigue strength is a material's resistance to repeated cycles of stress and/or heat. Most seats are made from either aluminum alloys or carbon-fiber composites, both of which have high fatigue strengths. A professional-level aluminum seat can easily cost $1,500 to $2,000. Carbon-fiber composite seats are stronger, lighter, and more energy absorbing, but they're also eight to ten times the cost of aluminum-alloy seats.

Once you've ensured that the driver is securely held in the car, it's up to the car to dissipate energy. The mass of the car and driver remain constant, so changing motion energy is mostly about changing velocity. Changing velocity requires a force.

When a driver goes from 180 mph to a stop, all of his (and his car's) motion energy has to be transformed into other types of energy. As a driver rolls to a stop in the pits, the brakes exert a force on the rotor. Motion energy is transformed into heat (friction between the brake pad and the rotor, or between the tires and the track) and possibly sound (squealing) or light (sparks or glowing rotors). The same thing happens when you approach a stop sign (hopefully without the squealing and glowing brake rotors). In a normal stop, energy is transformed in a controlled manner over a period of time, and you barely notice the force.

If you have to stop suddenly, you're more aware of the force because the change in velocity occurs more quickly. Crashes happen over a very short period of time, and instead of a smaller, sustained force, there is a force spike. Although the spike may only last a fraction of a second, the size of the spike can crumple quarter panels, twist steel, and worse. The size of the force spike depends on how much your velocity changes and how long the change in velocity takes.

Crush zones extend the time of collision by providing a sequential dissipation of energy. The tubing in the chassis gets smaller as you move away from the driver; the front and back bumpers deform more easily than the roll cage. The energy used to deform the front and back bumpers decreases the energy that reaches the driver. The new car design has stepped door bars for the same reason. The topmost horizontal bar sticks out farthest from the car. If the side of the car hits a wall, the top bar is pushed in first, then the next lower one, and so on. Each collapse dissipates energy.

It is important, however, that an impact not send pieces of the car flying. In July of 1999, NASCAR started requiring teams to use braided steel tethers on the front spindles (the shaft on which the wheel assembly rotates), rear deck lid, and hood. A tether is a cable with a loop at each end. One loop is connected to the potentially detachable car part and the other to the car's chassis. Steel was chosen for its strength; however, steel is heavy, so making stronger tethers by making them thicker or requiring multiple tethers wasn't an option.

In 1881, George Emery Goodfellow, a physician in Tombstone, Arizona, pulled a silk hanky from the breast pocket of a man injured in a gun battle and found two bullets. Although the man died from the force of the impact, there was not a single drop of blood on the handkerchief. The silk handkerchief had literally stopped the bullets. Spider silk, which is a natural polymer, turns out to be five times stronger than a piece of steel of the same weight.

We don't understand completely what makes spider silk so strong, but we have been able to make polymers with similarly exceptional strength for their weight. One of those polymers is Vectran, a liquid crystal polymer developed by Celanese Acetate. Vectran is comparable in strength to Kevlar, but Vectran is ten times more flexible, is resistant to chemicals, moisture, and abrasion, doesn't stretch as much, and doesn't weaken as much when repeatedly bent and pulled on. Vectran is used in blimps and sails, to reinforce rubber hoses and tires, and in the air bags that cushioned NASA's Pathfinder on its mission to Mars when it landed. NASCAR requires two Vectran tethers (in case one is severed) for each easily detachable part, including the rear wing.

Vectran tethers were first mandated in 2003, and Elliott Sadler was kind enough to test the new system at the second Talladega race that year. After tangling with Kurt Busch on the backstretch, Elliott's car flew into the air, flipped end-over-end, slid on its roof through the grass, and then did four or five barrel rolls before coming to rest. The tethers held.

So did Elliott, who walked away from the crash. These scary-looking crashes are actually a series of smaller velocity changes. Each time the car hits the ground, it loses a little velocity, and the driver feels a force proportional to that loss of velocity. Elliott felt a series of small forces instead of one large and potentially much more dangerous force. He probably didn't think they were small while it was happening, but aside from having the wind knocked out of him, he was uninjured.

Extending impact time decreases the size of the force spike. The simplest way to extend the time of an impact is to pad the driver's seat and helmet, and anything else the driver might hit inside the car. Race cars don't have air bags, but air bags are effective for the same reason. Instead of stopping suddenly, the air bag extends the time over which your velocity changes and thus decreases the maximum force you feel.

NASCAR works with researchers at Wayne State University, who use crash-test dummies outfitted with sensors that measure how different body parts move in a crash. The high energy scales of racing impacts can produce outcomes that might not be predicted from lower-speed passenger-car impacts.

The head and the torso are the most critical body parts to protect because they house your vital organs. The torso is held in the seat by the harnesses, which are designed to spread the impact over the strongest parts of the body. There's a reason they tell you to fasten your seatbelts low and tight across your hips—the pelvis is one of the strongest bone structures in your body. The shoulder harnesses are three inches wide, which spreads out the force instead of concentrating it in a narrow band.

NASCAR requires full-face helmets for drivers, as well as for any pit-crew members working with fuel, to protect against impacts and fire. The helmets have an outer shell made of a carbon-fiber and/or Kevlar-reinforced resin that is hard enough to prevent objects from penetrating. A foam layer inside absorbs and dissipates energy, and an inner Nomex lining provides protection from fire.

The driver's torso can be strapped tightly into the seat, but his head needs more freedom of movement. Too much freedom of movement, however, allows the driver's head to snap forward if he stops suddenly, and that type of injury can be fatal. Robert Hubbard, an engineering professor at Michigan State University, and his brother-in-law Jim Downing (a former race car driver), designed the HANS (Head and Neck Support) Device to prevent this type of injury. The HANS harness, which is made of a carbon-fiber-and-Kevlar composite, slips over the shoulders, runs around the back of the neck, and extends down either side of the chest like an upside-down "U." The shoulder harnesses hold the device onto the driver's torso. A rigid support, similar to what football players wear, comes up behind the neck.

Two flexible tethers about six inches long run from either side of the HANS device to the helmet. The tethers are normally slack enough to allow the driver to move his head freely. In an impact, the driver's head moves forward, and the tethers tighten before the seat belts stop the torso. This lets the head, neck, and torso continue to move forward as a single unit instead of having the neck and head snap forward relative to the torso.

The HANS device became available in 1991. Drivers complained that it was uncomfortable and too bulky, and that it would make it hard to get out of the car in case of an emergency. NASCAR felt it was important enough, however, that in October 2001, they started requiring that every driver use an approved head-and-neck restraint system.

A driver involved in an accident is required to take his HANS and his helmet with him to the infield care center. Tiny fractures in the HANS or the helmet may indicate that the impact was worse than it appeared. Even though a driver may seem fine, a cracked helmet may generate a referral for X-rays or an MRI scan to rule out internal injuries.

Having their own R&D center allows NASCAR to respond to current challenges and to anticipate new ones. Although they have made accidents survivable, they understand that they will never eliminate accidents entirely from motorsports.

"Look, if you give forty-three guys, mostly between the ages of twenty and thirty-five, cars this powerful," Mike Fisher said, "there's only so safe you can make it."

Fourteen
Making Racing SAFER

It was a fine January day in Charlotte. Unfortunately, neither I nor the team of engineers from the NASCAR R&D Center were in Charlotte. We were standing on an old airstrip just outside the Lincoln, Nebraska, airport, getting ready to watch a race car run into a concrete wall. The temperature was in the single digits, which prompted Dean Sicking, University of Nebraska–Lincoln professor of civil engineering and director of the Midwest Roadside Safety Facility (MwRSF), to suggest opening a branch campus: the University of Nebraska–Houston. Or anywhere warmer.

Dean still has just a hint of Texas in his voice, even though he's been at the University of Nebraska since 1992. The center he directs is nationally recognized for its development of highway safety devices like the boxes on the ends of guardrails that direct the rails downward instead of into the vehicle in the event of a crash. Although he is well recognized by his peers, most people know him for leading the development of the SAFER (Steel and Foam Energy Reduction) barriers that line every race track hosting a NASCAR race. In 2007, his work in highway and motorsports safety earned him the National Medal of Technology, the nation's highest honor for technology.

Today's test wasn't about walls: It was about the door structure of the new car. The car about to be sacrificed would be slammed driver's-side-door first into a concrete barrier. These are the most dangerous types of crashes because much of the car's energy dissipation comes from the front and rear bumpers, which don't provide much help in a side crash.

The media often cite the acceleration—the number of g's—measured by a car's black box when they report on a crash. On an earlier visit to the NASCAR R&D Center, I had asked Steve Peterson about the magnitudes of accelerations during particular crashes—and got another one of those pained looks. Steve went on to explain that a crash is a complex event that can't be described by a single number. Describing a crash by the peak number of g's experienced is like describing a snowflake by its diameter. Many other factors, like the time over which the impact takes place, the angle at which the car hits, and what part of the car hits first, are just as important.

Like snowflakes, you don't get to pick which types of crashes will occur. Each crash is a data point, but as Dean said, "You don't hope for data points." One of the most significant things that Steve Peterson has done is to create a database of NASCAR crash information, with as much information as can be gleaned from driver and witness reports, video evidence, and black-box data.

The severity of a crash depends on how fast the car is going when it hits the wall and the angle at which the car hits the wall. Researchers call the change in velocity Δv, which is read "delta v." The triangle (which is actually the Greek capital letter delta) indicates change.

The faster the car is moving, the more serious the crash is likely to be, but it's not a linear relationship. "Crash severity scales like the square of Δv," Dean explained, which means that hitting a wall twice as fast is four times worse. This dependence arises from the motion energy, which is proportional to the square of the speed.

The force felt in a crash is proportional to the change in velocity divided by how long it takes to change velocity, but you also have to take into account the direction you're going when you hit. Compare the two pictures below: The thick arrows indicate the speed of a car heading for a wall. The lengths of those arrows are the same, which tells you that the cars are going the same speed; however, the car on the left is approaching the barrier at a much greater angle than the car on the right.

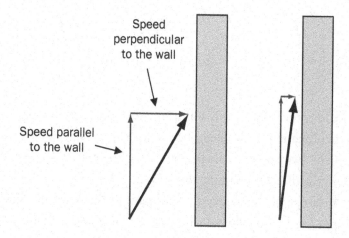

We can break the speed into two parts—one part parallel to the wall and one part perpendicular to the wall, which I've shown as thinner arrows. The speed perpendicular to the wall is much smaller when you approach it at a shallow angle. Even though the cars are going the same speed, more of that speed is directed into the wall when you approach it at a larger angle.

In the drawing on the next page, I've broken the velocities down into the parts parallel and perpendicular to the walls. Engineers call the speed in the perpendicular direction "lateral" and the speed in the direction parallel to the wall "longitudinal."

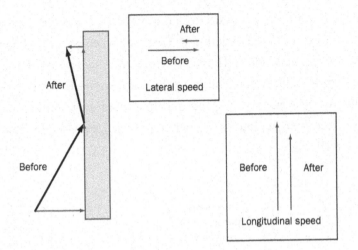

The longitudinal speed is a little less after the impact than before, but the direction is the same before and after. The difference be-tween the two speeds is small, indicating there wasn't a very large force in the direction parallel to the wall.

The change in lateral speed causes most of the problems. The lat-eral speed starts out toward the right, but is going to the left after the collision. Think about how much force you would need to stop a car coming at you compared to the force you'd need to not only stop the car, but push it back in the opposite direction. The force is much greater in the second case because there is a 180-degree change in direction. If you hit the wall at a shallow angle, the change in the lateral velocity is smaller because most of your speed is along the wall. If you hit the wall head on and bounce back (whether it's with the nose of the car or the driver's-side door), all of your speed is in the direction perpendicular to the wall, which is why these kinds of crashes are so serious.

There are two ways of making crashes safer: improving the walls and improving the cars. Although the test we were about to watch was of the car, Dean is best known for his work with walls. Given his research expertise, it is appropriate that Dean's involvement in motorsports came about accidentally.

The traditional barriers around race tracks were concrete walls and stacks of tires bundled together. The primary function of these structures was keeping the race cars on the track and not in the stands. They didn't provide much protection for the drivers. In 1998, the Indianapolis Motor Speedway installed a new type of barrier called the Polyethylene Energy Dissipation System (PEDS) in front of the existing concrete retaining wall. The PEDS had sixteen-inch-diameter high-density polyethylene (HDPE) tubes covered by an armor of inch-thick overlapping HDPE plates.

When a car hit the PEDS, some of the car's motion energy was used up moving the wall and deforming the tubes, leaving less energy to be dealt with by the driver. The first test of the new system was during an IROC race that year. The driver, Arie Luyendyk, suffered nothing more serious than a concussion, but the impact scattered debris from the wall all over the track.

"Luckily," Dean said, "Luyendyk was running fifth or sixth in a nine-car race," which meant there weren't many cars that had to dodge the debris.

Tony George, president of the Indy Racing League (IRL) and owner of the Indianapolis Motor Speedway, approached Dean's former employer, the Texas Transportation Institute at Texas A&M, to improve the PEDS system. The institute was already engaged in a similar project with a private company, so they recommended Dean.

The timing was bad. A law that required all new safety features on roadways to meet current federal guidelines had just been passed, which meant that the MwRSF had a lot of projects.

"We were very busy," Dean recalled, shaking his head. "All these ancient safety devices that states were using now had to be tested and shown to meet the guidelines, or scrapped and replacements developed. So we were flat overwhelmed with work." There were technical issues as well.

"One of the biggest problems was going to be figuring out how to make a race car go 175 mph without anyone in it," Dean explained. Not only would they have to develop the barrier, they'd have to develop entirely new procedures to test the barrier. Dean was on the verge of turning down the project, but Ron Faller, a research assistant professor at MwRSF, made an impassioned plea.

"Ron came to me and said, 'I want to do this.' Ron is an important cog in our facility, and he didn't come just representing himself. He had caucused with the other staff, and he told me it was unanimous that they wanted to do this work, so I relented and . . ."

I interrupted to ask why his researchers wanted to take on this new project so badly.

"They were race fans," Dean replied. "This was their opportunity to look inside IRL and, eventually, NASCAR." Dean thought this would be a one-time research project. They'd do the development, hand it off, and be out of racing. That was 1998.

There were a lot of constraints on the design. A very elastic wall would temporarily store much of the energy, but the car would be pushed back into traffic and hit by other cars. The barrier couldn't break into pieces that would be hazards for other cars. The surface of the wall couldn't grab the car and make it stop suddenly.

There also were practical issues. Putting a lot of effort into a barrier that would only work at one track didn't make sense. They wanted a design that could be modified for other tracks, including smaller tracks that didn't have much space for new walls. The wall couldn't be too expensive to install, and any repairs had to be easy and fast.

In addition to designing the barrier, they had to figure out how to make a car going 175 mph hit a wall at a well-defined angle.

"I actually didn't realize how interesting making a car go fast could become," Dean smiled. "Because of the expense of IRL [Indy Racing League] cars, they made us prove that we could make a car go fast, so we were testing with a Ford Festiva." The Festiva was the designated stand-in for an Indy car only because it was about the same weight.

"Have you ever seen a Ford Festiva go 110 miles per hour?" Dean laughed. "Their suspensions are really not built to handle that type of speed."

Dean laughed even more as he recalled their first test with a real IRL car, which took place in front of the IRL brass. They used a heavy-duty truck to get the car up to speed via an arrangement of pulleys and a steel tow rope.

"We get the car up to 90 miles per hour and there was so much drag in the Festiva that when the tow truck would shift gears, the Festiva would keep the tow cable taut." He used his hands to represent the tow truck and the race car. "A race car rolls better, so it gets to ninety, our tow truck shifts, and the race car catches up some with the tow cable and the tow cable drops off. We're at the point of no return, we can't brake the car, and we're coasting. It rolled into the wall at about eighty-five."

MwRSF now uses two tow trucks, one turbocharged and one with nitrous oxide, and a much different tow ratio—and they don't shift anymore. The ultimate goal had been 175 mph, but Dean says, "We towed an Indy car into a barrier at 150 miles per hour and 25 degrees. IRL said that was enough."

The barrier design started from the existing PEDS barrier. Although MwRSF improved the PEDS' performance, Dean knew that modifying the existing system wasn't going to be enough.

"We knew from the start that we had to get rid of the plastic skin," Dean said, in reference to the HDPE plates on the face of the barrier. "I knew we needed steel, but the racing industry's attitude was, 'Polymers are soft and steel is hard, so hitting polymers is better.' You can't just ignore your sponsors."

The HDPE front stayed, but MwRSF studied a range of energy-absorbing materials and configurations. By the time they had done all of the preliminary tests, Dean says, "We had convinced them that foam was the best energy-management system because it provided a much more controlled and consistent dissipation of energy."

In the late fall of 2000, the IRL administrators realized a potential problem: IRL wasn't the only sanctioning body using the Indianapolis Motor Speedway. Putting up a barrier for the IRL races and taking it down for NASCAR's Brickyard 400 wasn't very practical. Tony George called Bill France, Jr. to invite NASCAR to participate. NASCAR was already independently testing some alternative soft walls; however, they signed on to a cooperative venture with IRL. This partnership provided Dean's group with another challenge. Stock cars don't move as fast as IRL cars, but they are twice as heavy. When they ran a stock car into the HDPE/foam wall, the heavier stock car snagged in the plastic sheathing. The results were not encouraging.

"Basically, we proved we could be equivalent to concrete," Dean says. MwRSF convinced the sanctioning bodies to try using a steel plate with the same bending resistance as the HDPE armor and repeated the test. The results were much more promising.

"It was the first test where we had a significant reduction in the severity of the crash," Dean said. "We got comparable reduction in the lateral component of the acceleration pulse and there was no significant increase in the longitudinal pulse." In other words, the barrier behaved no worse than the concrete barrier in the direction along the wall, and significantly reduced the impact perpendicular to the wall.

By the end of 2000, MwRSF had demonstrated to the sanctioning bodies' satisfaction that steel was superior to plastic and were given the go-ahead to implement what would eventually become the

SAFER barrier. Then came Dale Earnhardt, Sr.'s death in February 2001. Although they had made huge progress, the system wasn't ready to be installed yet. Dean expressed frustration with the lack of recognition that the nature of research is to progress slowly, in bits and spurts.

"We run the DYNA model (a computer simulation program) for three days, and then we get an answer and we say, 'that's pretty good, but I think we could make it better by making this small design change,' and then we would turn it on and run it for three more days. Nothing happens with lightning speed like the race-car folks would like to see, and certainly not like the press would like to see."

The first SAFER barriers were constructed using two-inch steel tubing on the outside and rectangular stacks of foam between the steel and the existing concrete walls. Tests showed that the two-inch tubes were damaged too easily—something that hadn't been predicted by computer simulations. They changed to four-inch square tubes, which is the configuration that was installed at Indianapolis in 2002. Within ten days of installation, driver Robby McGehee spun into turn 3 during testing. Comparison of McGehee's crash into the SAFER barrier with a very similar crash by Eliseo Salazar a few weeks before the barrier had been installed showed that their new "soft walls" decreased the peak acceleration felt by McGehee by 60 percent. McGehee was back in the car a few days later.

The current version of the SAFER barriers uses square tubing eight inches on a side and 3/16" thick to form the impact plate that faces the track. The tubes are stacked and welded into twenty-eight-foot-long sections that curve with the track. Trapezoidal stacks of closed-cell polystyrene foam start out wide on the concrete wall side and get narrower as they approach the steel wall. A typical installation has about twenty inches between the two walls and polystyrene foam stacks every five and a half feet, but track constraints often require different configurations. The final components are high-strength nylon straps between the concrete and the metal tubing. When a car pushes in a section of the wall, the ends of that section try to flex outward. The straps anchor the ends of the section to the concrete wall. Nylon is used because of its ability to stretch and absorb the rebound energy without breaking or pulling out of the wall.

When a car hits a concrete barrier, the barrier doesn't move. All of the energy is absorbed by the car and its occupant. When a car hits a SAFER barrier, the tubular steel impact plate moves and the motion energy of the car is transformed into motion energy of the wall. The steel tubes transfer the load to the foam stacks, which compress and absorb energy. Instead of bouncing off the wall, the car is redirected at a shallow angle along the wall, which minimizes the chances of getting hit by oncoming traffic.

The SAFER barriers extend collision times from a tenth to two-tenths of a second, which may not seem like much, but it decreases force spikes by 30 to 80 percent, depending on the type of crash. NASCAR drivers have walked away from crashes with lateral Δv's as high as 80 mph—crashes that could have been fatal without the SAFER barriers.

Back at the Nebraska airstrip, there were no SAFER barriers to be seen: A plain, freshly painted white concrete wall was the target. They had crashed one of the old-style cars yesterday and a number of Dean's students were preparing today's test subject, a black, yellow, and white car that carried the MwRSF initials. Engineers from the NASCAR R&D Center and the University of Nebraska made final tests of the equipment installed in the car. A dummy with embedded sensors sat in the driver's seat. Four cameras focused on the interior of the car, and other sensors, including four NASCAR black boxes, were ready to measure the force of impact and the speed of the car when it hit.

The car was mounted on a sled attached to two I-beams. A single tow truck and a pulley system were configured so the truck could drive away from the wall while the car moved toward the wall, hopefully hitting at the target speed of 40 mph.

The impact would be videotaped with high-speed cameras to allow the engineers to analyze the crash down to the smallest detail. A gauge on the driver's-side door is connected to a flashbulb: When the car impacts the wall, the lightbulb flashes, allowing the engineers to identify the precise moment of contact when they review the videotape.

The engineers were finally satisfied that they had checked and double-checked all of the equipment. They would only get one shot at getting it right. In addition to the people working on the test, a couple of other engineering students had come to watch. We joked about the cold through clenched teeth as we moved toward the side of the enclosure. The smart folks ducked into a shed to escape the wind.

Steve Peterson sat on a stack of containers about ten feet up, out of the camera's view, but in perfect position to see the moment of impact. He didn't complain about the weather, just sat cross-legged on the containers, intent on the impending test.

There was no announcement, no warning when the test was about to start. The tires on the pickup smoked. The car started moving and slowly accelerated. Even though I was watching the car move toward the wall and I knew it was about to hit, the dull thud of contact made me jump. The car hit the wall so hard that you could read the tire code from the imprint the tires made on the wall. The students who were here to see "a crash" looked a little disappointed. It wasn't spectacular, but to the engineers, the interesting part had just started.

If this had been a crash during a race, the car would be headed for the hauler. The right-front wheel—and remember that the car hit on its *left* side—was badly dislocated, which the engineers would learn was due to a broken tie rod when they opened the hood. Three or four people took pictures from all possible angles. Technicians retrieved the digital movie cameras, which collect 500 frames per second, from the car and retreated into the heated trailer.

Steve walked around the car, eyeing it like a detective surveys a crime scene. He called Dean over to point out specifics of how the car had deformed during the crash. The pins were pulled from the hood and the hood raised. Peterson removed a crumpled crush panel and tossed it off to one side. That piece, along with the rest of the car, would be sent back to the NASCAR R&D Center for further analysis.

Once Steve was satisfied that enough pictures had been taken, Sawzalls and metal shears arrived and the engineers removed the door skin. The driver's-side door is protected by a 90-mil steel plate sitting just outside the door bars. The right-side door is reinforced similarly with a panel made from Tegris, the same material from which the splitter is made. Between these reinforcing plates and the door skins are blocks of energy-absorbing foam.

The NASCAR R&D Center tested more than 200 different foams before selecting IMPAXX, a closed-cell low-density polystyrene foam made by Dow. Polystyrene is a polymer containing hydrogen and carbon atoms that is extruded and made into a foam by heating the polymer until it liquefies and then using carbon-dioxide gas to blow the foam into shape. The gas creates holes, much like the holes in Swiss cheese. The best-known polystyrene foam is STYROFOAM Brand Foam, which also was invented by Dow.

The holes in open-cell foams are interconnected, which allows air and fluids to pass through them and makes them flexible. The holes in closed-cell foams like IMPAXX are not interconnected.

Each hole is surrounded by polystyrene walls. Open-cell foams are squishy, while closed-cell foams are rigid. The foam pieces used to protect electronics in their shipping boxes are usually closed-cell polystyrene.

When you crush a piece of foam, the gas inside the pores is compressed. Open-cell foams crush faster and more easily because the gas can escape through the interconnected pores, and the structure is less rigid. Closed-cell foams offer much more resistance. The stiffness of the foam is determined by the strength of the polymer walls and the size and density of the pores. IMPAXX foam deforms in stages: It compresses first and then buckles. Both processes dissipate energy.

Although the SAFER barriers and the new car design are making racing much safer, there are still problems to solve. Foam energy absorbers, like those used in the SAFER barriers, tend to be less energy-absorbing with each subsequent hit, and they are brittle when hit at high speed. Battelle has developed a new hyperelastic foam called FlexAll that is incredibly resilient, even under high-speed impacts. FlexAll is a closed-cell polyurethane. Arranged in a honeycomb pattern, it works well at the ends of the pit walls, which are currently protected by sand- or water-filled containers. The foam stretches or compresses as much as seven times its original size, but then decompresses in just a few minutes. It stands up to high-speed hits and doesn't lose its energy-absorbing ability after repeated impacts.

At the test site, Dean and Steve continued to stand in the cold talking excitedly about the differences between today's results and yesterday's results with the old car. It will take them a couple of weeks to analyze and understand all of the data they collected in the few short moments it took to turn a perfectly good race car into a pile of scrap metal. This car will never race again, but its sacrifice will help ensure that NASCAR drivers get to enjoy growing old.

Fifteen

Practice Doesn't Always Make Perfect

Martinsville Speedway is so far south in Virginia that it's almost in North Carolina. It was the last week of March and the trees were just starting to show a little green. A little before 7:00 A.M. (garage-opening time this Saturday morning), the sun was rising and burning off the large pockets of fog sitting in the valleys around the race track.

This was my second weekend with the No. 19 team. I had arrived Friday morning to a hearty, "Hey, you're back!" from Chris Miko, who took a few moments to explain how this visit would compare to my experience in Atlanta. Martinsville is a very different race track than Atlanta—physically and philosophically. The metal bleachers surrounding the track, which is a little more than a half-mile around, hold just 91,000 fans. Chris explained that there is no camping inside the track because there is no room. Even the drivers have to park their RVs outside.

Short-track racing is an entirely different animal than aero-racing in Atlanta. Short-track races are usually about mechanical grip and brakes. This race is also the second-ever race using what we were then calling the "Car of Tomorrow," a moniker outdated now that it is the only car being used. Chad Johnston, the team's engineer, was still stewing over the team's first experience with the new car at Bristol last week. A wheel vibration forced a green-flag pit stop that took them from leading to being down a couple of laps. They got a 27th-place finish instead of what should have been at least a top-5.

There were to be two practices today, at 10:00 A.M. and 12:50 P.M. The latter of these (ironically called "Happy Hour") is the last chance the team has to see what the car does on the track, although they can continue to work until the garage closes at 3:30 P.M. The No. 19 team qualified 18th yesterday, so there is work to do.

In the hauler's command center, Chad reviewed a video that superposes Elliott's qualifying lap with that of Denny Hamlin's No. 11, which won the pole.

"See," Chad said, pointing at the screen, "He wasn't getting off turn 2 as quickly. That's where he lost time." Elliott was a little more than two-tenths of a second slower than the pole winner. Chad pointed to the space on his computer screen between the No. 11 and the No. 19 as Hamlin's car crossed the start-finish line. The distance is less than a car length.

"There are sixteen cars in that gap," Chad said.

Elliott, already in his firesuit and a Dodge baseball cap, checked in at about 9:20 A.M. and sprawled out on the bench near the door of the command center. The videotape didn't tell him anything he hadn't known immediately after his qualifying run: They would need to use today's practice to figure out how to get out of turn 2 faster.

Kirk Almquist, the car director ("Head mechanic is what that means," he explained), appeared at the door. Elliott, whose long legs were blocking the entrance, looked up with a smile that made it clear he didn't plan to move. Kirk made some exaggerated noises of exasperation and climbed over Elliott to sit beside him on the bench.

Twenty-eight-year-old Kirk and his twin brother Clint grew up in California. They've been racing most of their lives. They decided one day to move to North Carolina and see if they could break into NASCAR. After working their way up through the other series, Kirk came to Gillett Evernham from Richard Childress Racing, which is where Clint still works on the No. 07 car.

Kirk is a stickler for details—a good trait for a car director (or car chief, as other teams call it). You can ask him anything about the car—new or old version—and he can tell you the answer without looking at the NASCAR rulebook that is always present in his back pocket. During the switchover from race to qualifying trim yesterday, Elliott was taping the right side of the car's grille while Kirk did the left. As soon as Kirk was sure Elliott wasn't looking, Kirk went back and trimmed Elliott's side. Kirk has a strong jaw and an almost military bearing. He's usually very serious, but he was in a rambunctious mood this morning.

"The problem," Elliott was saying, "is that the car just won't turn."

"Well, maybe you should start working out," Kirk volunteered. A short wrestling match broke out, but the room was too small for it to go on very long. A few moments later, Elliott was talking about steering boxes.

"If the eight-to-one works, I can put it on my Phoenix car," he said.

"*My* car? You mean *our* car," Kirk chided. "Me, me, me. You only think of yourself." Another brief round of wrestling ensued before Kirk asked Josh for the changes to be made before the practice that was to start in thirty minutes.

"Let's change the rear gear and the brake calipers and the transmission," Elliott suggested. Kirk rolled his eyes.

"And move the engine down an eighth of an inch," Chad said

very seriously. Josh looked on, grinning, before relaying the real instructions.

The garage at Martinsville is an aluminum roof, steel column supports, and a cinder-block half-wall at the back. The rear of the garage is about ten feet from the backstretch, so when cars are on the track, you might as well be standing on pit road during a race because it's so loud. The good thing, says Kiwi, the shock specialist, is that the garage stalls at Martinsville are large. At some tracks, he says, you can't lie down under the car without getting stepped on.

I was surprised by how close together the cars are in the garage. Only about ten feet separated the right side of our car from the No. 20 car next to us. Toolboxes line the backs of the stalls. I watched the proceedings from the four-foot space between our toolbox and that of the No. 43 on our left.

Kirk supervises three mechanics and an engine tuner. Allen Mincey, a thirty-five-year-old native of St. Augustine, Florida, works mostly on the rear of the car. Twenty-nine-year-old Tony Lunders from Washington State works primarily on the front of the car, and Ramon "Razor" Zambrano, a twenty-six-year old originally from El Salvador but now a North Carolinian, is responsible for the car's interior. Tom Engleson, the twenty-seven-year old engine tuner, is from Minnesota. During a slow moment, Allen told me that he likes this team because they work well together. Everyone knows what to do and does it.

Engines and aerodynamics are critical at tracks like Atlanta, but short tracks like Martinsville demand a lot from the brakes. Drivers come into the corners fast and have to brake hard to make the turns. The car's disc brakes slow the car by friction between the rotor (the disc) and the brake pads. Stopping a 3,600 pound race car and driver over a distance of 300 feet from a speed of 120 mph requires a force of 5,772 pounds. Stopping from 40 mph over the same distance requires a force of 640 pounds, but you don't provide all that force yourself.

The brake pedal is a lever that provides the same type of mechanical advantage as a seesaw. A lighter person can lift a much heavier person on a seesaw if the heavier person is closer to the pivot and the lighter person is further away. The seesaw multiplies the force the lighter person can exert, although there's a trade-off: The lighter person has to move a greater distance up or down than the heavier one. The brake pedal similarly multiplies the force you apply, usually by a factor of between three and six, but the brake pedal moves much farther than the brake pads do.

A second assist comes from the brake fluid. Squeezing a liquid doesn't change its volume significantly, which allows liquids to transmit forces. Place movable pistons at either end of a tube of liquid. Pressing on the piston at one end will move the piston on the other end. An advantage of using fluids is that the tubes containing them can twist and turn, so fluids can transmit forces over paths impossible to access with a rigid piece of metal.

Fluid pressure is the same throughout the tube. The force exerted by the fluid is the pressure times the piston area, so if the pistons on either end of the brake line have the same area, the force applied on one end will be the same as the force produced on the other. You can increase or decrease the force by making the piston areas larger or smaller. If the piston pushed by the brake pedal has a smaller area than the piston on the other end, the force exerted on the far end will be larger than the force from the brake pedal. The penalty is that the larger-area piston doesn't move as far as the smaller-area piston.

The brake pedal activates two cylinders of fluid, one for the front and one for the rear brakes. The force is normally split equally between the front and the back, but a brake bias knob inside the cockpit allows the driver to increase the proportion of the force going to the front or rear brakes. Adding rear brake can help a tight car enter turns better.

The force originating at the break pedal is transmitted to sets of pistons located in the brake calipers, which also house the brake pads. The brakes used at Martinsville have six pistons in each caliper, but brakes for tracks that don't require heavy braking may only have four pistons per caliper. The rear brakes only have four calipers, but they are larger here than they would be at an intermediate track. Racing brakes have fixed calipers, which have pistons on both sides of the rotor. Passenger cars normally use floating calipers, which have a stationary brake pad on one side and a piston-actuated brake pad on the other.

The pistons press the brake pads against the rotor and two types of friction slow the car. Abrasive friction is caused by the rotors abrading the brake pads. Motion energy is used to break bonds between atoms in the brake pad. The second type of friction is adherent friction. Some of the brake pad material comes off and forms a thin film on the rotor. The brake pad and the thin film on the rotor repeatedly form and break bonds, sort of like walking through something sticky. Adherent friction is one reason a track gets more grip as more rubber is deposited on it.

Tony carried a pair of front rotors, one in each hand, to the garage. I could tell from the way he was walking that the discs are heavy. They are a little less than a foot in diameter, about an inch and a half thick, and made of gray cast iron. Rotors used at other tracks are thinner and lighter, but the sturdy rotor they're using today is necessary to dissipate the heat that will be generated by Elliott's constant brake use.

The brake rotors aren't solid disks. They look like a sandwich: two thinner disks separated by metal vanes that allow air to flow through and dissipate heat. The airflow is assisted by brake fans and ducts, which move air from the square holes in the front bumper to the rotor's eye.

NASCAR allowed teams an extra brake duct for this race because of concerns about overheating. If the rotors get much hotter than 1,200°F to 1,300°F, small cementite inclusions can form. Cementite is abrasive, which wears the pad faster, but it's also a poor thermal conductor, which leads to uneven heating that stresses the rotor and can make it crack.

Formula One cars use reinforced carbon-carbon composite rotors, which are made from the same material used for the nose cone on the space shuttle. Carbon-carbon composites are graphite reinforced with carbon fibers, and they work at temperatures of more than 3000°F. The rotors are made by shaping carbon-fiber filament or cloth and then surrounding it with an organic binder. The binder decomposes when heated, leaving behind relatively pure carbon. A carbon-forming gas forced into the piece at high temperature over several days fills voids and forms larger graphite crystals. The complexity and time required is why a carbon rotor may cost ten times more than a cast iron rotor. Carbon rotors generally don't work as well at lower temperatures, but their cost is why NASCAR mandates cast iron.

Racing brake pads are made of a metallic-carbon composite. Abrasive particles in the pad provide friction, polymers make the pad less brittle, fibrous components hold the pad together, and metal particles help the pad dissipate heat. As with tires, there are trade-offs between friction and longevity. The amount of grip can change significantly with temperature. Brake pads work best in a narrow temperature window determined by the pad's composition. Stock-car pads usually operate best around 1,100°F.

At the superspeedways, drivers use their brakes only during pit stops or emergencies, so the brakes don't heat up as much. Brake pads used at those tracks have to have good grip even when they are relatively cool. At Martinsville, the brake pads stay hot the entire race and the pads have to work reliably at higher temperatures.

Tony showed me three stripes of temperature-sensitive paint running across the edges of the front rotors he had just removed from the car. The paints oxidize, which turns them whitish, when they reach a particular temperature. You can get paints that change color at virtually any temperature, but for racing applications, you usually want one around 800°F, one near 1,000°F, and one just above 1,400°F. The lowest-temperature paint should oxidize fully, the middle one partially, and the highest one not at all, which indicates that the brakes are getting hot enough to work, but not hot enough to thermally stress the rotor.

Although much of the brake heat is dissipated in the rotor, some heat is radiated to the surrounding parts, including the wheel. Outside the hauler, Swifty, the tire specialist, was armed with rolls of silver reflective tape. He explained an experiment he tried earlier this week after the No. 10 car melted a tire bead last week at Bristol.

"I took a torch and held it on the wheel for twenty seconds," Swifty said, "and I measured the temperature of the wheel at the bead. It was 256°F." The bead, he said, will melt at about 400°F, but it softens at even lower temperatures.

"I put one layer of tape on the wheel and the temperature went down to 188°F," he continued, "and two layers took it down to 156°F." He decided that one layer of tape was enough, so he and Ramon were taping the front wheels for today's practice. They attached the reflective tape carefully so that it couldn't come loose and get stuck in the brakes.

The track went hot at 9:55 A.M. On about the seventh lap, Josh radioed to Elliott, "We're the best car on the track." A half-dozen laps later, Josh asked Elliott to come in for a sway-bar change. Brett, Elliott's spotter, told Josh that the car looked good in turn 4, but still needed a little help off turn 2.

"You've got a rocket in three and four, man," he told Elliott from his vantage point near the top of the grandstands.

The team had moved up in the garage this week because garage spots were now being assigned using the 2007 owner's points. The No. 19 was in 13th place, between the No. 20 and the No. 43. The team had become more circumspect about their setup. They didn't say as much over the radio and relied more on written notes. Josh's changes were made quickly and Elliott headed back out onto the track.

"Steerin' box is starting to mess up," Elliott said over the radio. Josh told Elliott to come in right away. Before they changed the box, Josh insisted on one more run, replacing the new sway bar with the original one to make sure the sway bar hadn't caused the problem. After a few laps with the old sway bar, Elliott said that it was "definitely the steering box," but that it could wait until after practice to be changed.

After a longer run, Josh called for a plug check. Elliott revved the engine, and then shut it off and coasted onto pit road. Tom, the engine tuner, was waiting on pit road and extracted the spark plugs to "read" them. He was looking for things like oil, which would indicate that the piston rings are not sealing properly, and at the colors of the deposits on the ceramic around the spark plug. If there is a black sooty buildup, for example, Tom knows to check the fuel/air ratio because the carbon in the fuel isn't combusting completely.

The spark plug uses electrical energy from the battery to initiate the combustion event. The spark is generated by the same phenomenon that zaps you when you reach out for a metal doorknob in winter. When you shuffle along a carpeted floor, the carpet rubs electrons off you. This leaves the carpet with extra electrons, making it negatively charged, and you with missing electrons, so you become positively charged. The same charges are there, but they are distributed differently.

Charged objects tend to try to make themselves neutral. A positively charged object will try to attract electrons and a negatively charged object will try to get rid of electrons. When you reach out to the metal doorknob, electrons in the doorknob move toward your positively charged hand. Metal is a good electrical conductor, so it is easy for the electrons in the metal to move. Air, however, is not a good electrical conductor. When enough electrons build up at the surface of the doorknob, they can jump from the doorknob to your hand. That's what the zap is: about a million million electrons jumping through the air from the doorknob to you. A bigger zap means more electrons are involved. Lightning works the same way, but with about a hundred million more electrons than your average doorknob zap. Charges build up in the cloud until there are enough for them to jump from the cloud to the ground. The peak power of an average lightning bolt is equal to 10 billion 100-watt lightbulbs.

A spark plug generates a separation of charge between the ground—the L-shaped piece of metal protruding from the top—and the electrode, which is a metallic pin at the spark plug's center. A ceramic insulator (the part that breaks when you overtighten the spark plug) separates the two so that the only way for electrons to flow between them is through the air gap.

A spark plug produces a mini–lightning bolt. Electrons build up at the tip of the electrode. They'd like to get away from each other, but they can't because of the air gap. If enough build up, the electrons jump across the gap between the electrode and the ground. The voltage at the spark plug gap can be anywhere from 40,000 to 100,000 volts. One of Tom's jobs is to adjust the gap in the spark plugs: If the gap is too small, the electrons jump before they have built up enough charge. If the gap is too large, a spark may not be generated at all.

We finished first practice in 15th place and the crew prepared to change the steering box. The steering box contains gears that translate turns of the steering wheel into motion of the wheels. A 12:1 steering box means that the output (which goes to the wheels) rotates once when the input shaft (from the steering wheel) turns twelve times. A lower steering ratio gives a quicker response. The crew also changed the power steering pump. The pump pressurizes a fluid that adds to the force you exert on the steering wheel. You have to turn the steering wheel the same amount, but the power steering reduces how much effort it requires.

It was past noon, but the crew didn't take a lunch break. Tony was lying under the car with a wrench in one hand and a sandwich in the other. I realized the real utility of the new car's front splitter as I watched the crew change the steering box: The splitter held five lug nuts, a roll of tape, two bottles of water, a razor blade, a screwdriver, and a plate.

The last thing the crew does each time they change the suspension is to scale the car. They pulled out four scales, each of which is about two feet long and a foot wide, with a solid platform immediately in front of the scale. Tape marks on the garage floor ensure that the scales are set in the same places each time. Allen jacked up the right side of the car and Ramon the left so that Tony and Kirk could slide one scale under each wheel.

Cables run from each scale to a readout that sits in a small metal suitcase. The high sides of the suitcase provide storage for the cables and keep anyone else from seeing the numbers. The car is lowered onto the solid platforms and then rolled back onto the scales. Allen bounced the rear end up and down and Kirk did the same in the front. Allen explained that this settles the car on the springs. Kirk looked at the readout and asked Allen to change the left-side weight.

NASCAR requires a minimum weight of 3,450 pounds, with at least 1,700 pounds of that on the right side. The old car had a minimum weight of 3,400 pounds and required 1,625 pounds on the right side. Weight shifts to the outside of the car in the turns, and on NASCAR's oval tracks, that means to the right. Without a minimum right-side weight, teams would shift a lot of weight to the left to help keep the left-side tires on the track. An overly heavy left side, however, would make it difficult for the driver to control the car should he have to steer sharply to the right.

The ballast—extra weight used to adjust the weight distribution—goes in the chassis frame rails. While Allen is unscrewing a cover plate on the rear of the left frame rails, Kirk waves me over and points to two painted metal bricks, each of which is about three inches by four inches by six inches.

"One is tungsten," he said, pointing, "and the other is lead. Which one do you think is heavier?"

The periodic table of elements shows that tungsten has atomic number 74 and lead has atomic number 82. The atomic number tells you the number of protons in the nucleus and the number of electrons in the atom. An atom with a higher atomic number is heavier than an atom with a lower atomic number, so I guessed lead.

Kirk smiled, so I knew I was wrong before I even tried lifting the two blocks. The lead block is about thirty pounds. The tungsten block is about fifty pounds. Lead atoms are heavier than tungsten atoms, but tungsten atoms pack together more closely than lead atoms. Even though the two pieces of metal are the same size, there are more tungsten atoms than lead atoms. Allen, who had been watching with amusement, asked us if we were done with his ballast, and then slid the tungsten weight into the frame rail, followed by aluminum tubing to keep the weight from sliding. He screwed the cover back on and duct taped it for extra security.

After re-scaling the car, Kirk asked Allen to add a round of "wedge." Wedge (also called crossweight) is a percentage of the total weight: the weight on the right front plus the weight on the left rear divided by the total weight. If the car's center of gravity is at its geometrical center, using four equal springs will put one-quarter of the car's weight on each spring. The grip is proportional to the force pushing down on the tire, and changing the spring preload—the amount by which the spring is initially compressed—changes how the weight is distributed among the four tires.

Adding wedge means increasing the weight on the left rear and the right front, and for a car turning left, increasing wedge tends to make the car tighter. Allen placed a long socket wrench into the hole in the left side of the rear window. When he turns the wrench, the plate holding the spring moves down and compresses the spring. This increases the weight not only on that wheel, but on the diagonal wheel as well, so the weight on the right front also increases. The hole in the left-hand side of the rear window is for adjusting the preload on the left rear spring. They have to open the hood to adjust the preload on the front springs.

There are two holes in the right side of the rear window. The top one is for adjusting wedge and the one below it is for adjusting the trackbar. The trackbar runs from the back of the left-side trailing arm to the frame on the right. Unlike the front suspension, in which the two wheels move independently, the rear wheels move as a unit. The truck arms locate the rear axle front-to-back and allow it to move up and down, and the trackbar controls left-right motion. Moving the trackbar up pulls the wheels to the right and moving it down moves the wheels to the left. Moving the wheels to the right increases how much the car rolls in the turns. Raising the trackbar makes the car looser and lowering it makes the car tighter.

The final step in getting the car ready for Happy Hour is checking the wheel alignment. The toe is the orientation of the wheels when you look at the car head-on. If the wheels are straight up and down, there is no toe. If the fronts of the tires are farther apart than the backs of the tires, the car is "toed out." "Toed in" is the opposite: The fronts of the tires are closer together than the backs of the tires, making the car look pigeon-toed. Kirk, who was kneeling by the right-front tire, slid a tape measure under the car to Tony, who was stationed at the left-front tire. Both placed the tape measure in slotted aluminum plates they had stood vertically against the wheels, which let them measure the distance between the tires.

Tony then measured the camber, which is the tilt of the front wheels with respect to vertical. Zero camber corresponds to the wheel being straight up and down. Negative camber is having the top of the wheel closer to the car's centerline than the bottom of the wheel. Camber helps compensate for the changing angle of the wheel as it travels on the banking. Cars for oval tracks have positive camber in the left front and negative camber in the right front.

I realized it was getting close to Happy Hour when Tony Stewart got into his car and started gunning the engine. The fact that the car was on jack stands and didn't have any tires didn't seem to dampen his enthusiasm. The No. 20 crew ignored him and continued about their business. Stewart had just pulled out to get in line for practice as the last lug nuts on the No. 19 were being tightened. Elliott climbed into the car after a brief consultation with Josh.

After just a few minutes on the track, Elliott came over the radio.

"Don't know if the track has slowed down," he said, sounding frustrated. "I just can't turn." By the time I had looked up from my notes, the car was back in the garage. Elliott asked if the box was an 8:1 box and said it felt more like a 12:1 because the car wasn't as responsive as he expected it to be.

After verifying that it was, indeed, an 8:1 steering box, Josh and Chad stopped a moment to talk with Kirk. This wasn't the problem they had expected to be dealing with during this practice. After a few minutes they decided to change the right-side control arms. Elliott got out of the car and watched the other cars' lap times on the computer monitor hanging from the toolbox with a concerned look. Time became an issue. The control arms were not cooperating. Tony, who is a slender five-foot-nine, was standing on the splitter, using his whole body to pull on a wrench. Kirk came over to add his muscle and the recalcitrant control arm finally yielded. Elliott took the car out on the track again while the crew stood in the empty garage, waiting.

"It's real bad," Elliott reported.

"Ten-four," Josh replied from on top of the hauler, "I can see it." Kirk grimaced and kicked at a piece of tape on the floor as the crew started grabbing tools. Josh and Chad decided to undo the control arm adjustment they had just made. They also changed two springs, which meant they had to re-scale the car. Kiwi took the springs that were removed and taped over the engraved spring rate to prevent advertising what we were running.

Difficult times are when the No. 19's teamwork becomes most noticeable. Kirk and Tony struggled again with the obstinate control arm. Allen peered over their shoulders and, without saying anything, grabbed a pry bar from the toolbox and handed it to Kirk. Kirk took it as if he had just asked Allen to hand it to him. With thirty minutes left, Elliott headed back to the track.

"This will not turn at all," Elliott said disgustedly, "Zero turnability." Kirk exhaled loudly and shook his head.

"I'm afraid we're chasing the steering box," Josh said. At least he and Chad were getting a workout going up and down the stairs to the hauler roof.

Elliott came in for one more change. The crew waited anxiously, but with decidedly less optimism.

"No different," Elliott said after a few moments back on the track.

"Okay," Josh said disgustedly. "We're done, guys, we're done." We ended up "forty-first quick," as the broadcaster Larry McReynolds would say, or "third slowest," as Chad said. The garage would close in eighty minutes. While Josh was meeting with the other team directors, the crew made a discovery.

Somewhere along the way, a shim—a thin piece of metal—was replaced, except instead of taking out the old shim and putting in a new one, the new shim was added on top of the old one. Josh returned and was informed, but he had no way to know whether that was the reason for the ill-handling car, or whether there was still a problem with the steering box. He asked the crew to put the old steering box back and gave Kirk a list of changes based on what seemed to be working for the No. 9 and the No. 10. How the car behaves tomorrow will be a surprise to everyone.

We all hope it will be the good kind.

Sixteen

Elliott Is Hot

Cars were pouring into the Martinsville parking lots by 6:15 Sunday morning, even though the gates didn't open until eight. The garage wouldn't open until seven, so I sat in my car watching the sun rise and listening to a preacher and his companion sing hymns with a small group gathered around the No. 31 souvenir hauler.

A van pulled up in the row facing me, and then a car next to it. A mom and dad, one set of grandparents, and three kids emerged from the vehicles. They set up two canopies and the men put up a flagpole with an "8" flag on top and a "1" flag underneath. The mom helped the kids start the coffeepot. A full breakfast was ready less than thirty minutes after unloading.

Down the hill near the garage entrance, a couple of hundred crew members were milling about. At seven sharp, the gates opened and a flood of colorful uniforms passed through the gates, over the pit wall, and into the garage.

Among those crew members was Ramon "Razor" Zambrano. You're most likely to see him when the team is pushing the car somewhere—he's the one reaching in the driver's window and steering. Twenty-six-year-old Ramon, who emigrated to the U.S. from El Salvador in 1999, is the youngest member of the garage crew. He started with Gillett Evernham Motorsports on the No. 9 car and moved to the No. 19 last year.

Crew members were in the hauler getting breakfast or coffee and commiserating about how hard it is to keep in shape on the road. Ramon, who is about five-foot-nine with high cheekbones and wavy brown hair, volunteered that although he eats as much as he can, he can't seem to get his weight over 145 pounds. This elicited absolutely zero sympathy from any of us.

Ramon's primary responsibility is the inside of the car, which includes maintaining and preparing Elliott's safety equipment. Although the weather was a little chilly, Elliott would have no problem staying warm today because the driver's cockpit can reach 120°F to 140°F during a race, thanks to that 850 horsepower engine under the hood. Radiated heat warms the metal firewall and floorboard. Aerodynamic concerns limit the amount of airflow through the car, and the cockpit is not air conditioned because it would use engine power that could be going to the wheels.

Ramon explained that they put thermal blankets on the floorboard underneath Elliott's feet. The ceramic insulating material in the blankets (which was developed for the space shuttle) is strong and flexible to 2,000°F, and can even handle 3,000°F for short stretches. A half-inch layer of heat shield can make the cockpit 40 percent cooler. In one test, the transmission tunnel reached 170°F after thirty-two laps without insulation, but only 110°F after sixty-four laps with insulation.

Elliott also wears fire-resistant socks and shoes. Driving shoes are special not just in providing heat protection, but in having low-cut edges on their rubber soles to prevent than from catching on the pedals. The rolled heels make good pivots for quick pedal changes. Elliott adds heat shields that wrap around his heels. Their silver outsides reflect heat the same way a mirror reflects light. The insides of the heat shields are fire-resistant and thermally insulating.

Elliott's primary protection is his bright red-and-white firesuit, which covers him from neck to ankles. Knitted cuffs hold the firesuit closed at the wrists and ankles, and a Velcro closure secures the neck. Fire-retardant gloves finish off the ensemble, usually with a fire-resistant leather or other texturized inset in the palm to help the driver grip the wheel.

Firesuits are quilted, which initially seems a little odd because we associate quilts with staying warm. Quilting traps air between layers of fabric, and that trapped air is a good thermal insulator. When he's standing in front of the hauler talking to the media, Elliott's skin is warmer than the outside air, and the air pockets slow the escape of his body heat. When he's in the car, the air outside his firesuit is warmer than the air near his skin, and the quilting minimizes how much of the warmer air penetrates the firesuit. Firesuits should have a relaxed fit because heat reaches the skin faster when the skin is in direct contact with the fabric.

Your body has its own cooling system, with your skin serving as the radiator. Warm blood moves to the skin's surface through many tiny capillaries. The large surface area of the capillaries helps transfer heat from the blood to the surrounding air, just like the fins on a radiator.

When it gets warm, you can't get rid of heat fast enough through blood flow and you start to sweat. Evaporation—changing a liquid to a vapor—requires adding heat, just as it takes heat to turn water to steam. Turning sweat into vapor pulls heat from the surrounding air, which makes you cooler.

Sweat shouldn't collect inside the firesuit. In addition to being uncomfortable, heat can turn sweat into steam, and steam trapped close to the skin can cause second- or third-degree burns. Firesuits "wick" moisture away from the skin. The force between a fabric molecule and a sweat molecule is stronger than the force between two sweat molecules, and this draws sweat molecules to the fabric and away from the skin. Paper towels work the same way.

Some sweating is good, because it helps the body cool down; however, too much sweating leads to dehydration. Pit crews used to pass paper cups of water to the driver during pit stops, but Elliott enjoys a much better system. Ramon poured a diluted sports drink and ice into a shiny silver thermally insulated pouch that he then secured in the car. Plastic tubing runs from the pouch to the driver's mouth. A small battery-operated pump allows the liquid to flow when the driver bites on the end of the tube.

Keeping the driver cool and hydrated is much more than a comfort issue. Heat can cause mental and physical exhaustion and promote "irritability, anger, and other emotional states that may result in rash or careless behaviour," according to a study done at the University of Western Australia. A driver who loses more than 3 percent of his body weight in fluids is at risk for fatigue-induced errors. A 3-percent fluid loss for Elliott, who weighs 195 pounds, is about six pounds. A driver can sweat the equivalent of ten pounds of body weight during a race, which is comparable to the fluid loss of marathon runners and football players.

Severe dehydration can cause much more serious problems than crabby drivers. A dehydrated driver can't sweat, which leads to a rapid increase in body temperature. Blood pressure drops, which makes heart and respiration rates increase, which in turn increases body temperature even more. Body temperatures above 104°F (40°C) are life threatening.

Although firesuits protect against heat, the most important reason for wearing a firesuit is, of course, the threat of fire. Elliott Sadler said, "Two fears you have as a race-car driver: One is being on fire, and two is being T-boned in the driver door—everything else you sort of accept." If you break a bone, or are even knocked unconscious, emergency personnel can come to your rescue.

Fire requires three things: Heat, fuel, and oxygen. When you light a candle, heat from the match vaporizes the candle wax, which is the fuel. The vaporized wax molecules react with oxygen in the air to produce heat, and that heat vaporizes more fuel. Fire is difficult to stop because each step feeds the next. A typical candle flame burns at about 1,400°F, but a gasoline fire burns between 1,800°F and 2,100°F. Methanol fires can reach almost 3,500°F.

"Firesuit" is a bit of a misnomer: It is technically a fire-*retardant* suit. There is no such thing as fire*proof.* Any material exposed to enough heat for enough time will either burn or melt. The idea is to put as much distance and time as possible between the fire and the driver.

Among the many patches on Elliott's firesuit is a black-and-white patch that says "SFI 3.2A/5." SFI is a nonprofit organization that develops specifications for racing equipment, many of which are incorporated into NASCAR rules. 3.2A is the specification for firesuits. Elliott and anyone on his crew handling gasoline during the race must wear a SFI 3.2A/5 firesuit. The rest of Elliott's over-the-wall pit-crew members must wear at least SFI 3.2A/1–rated firesuits.

A larger number after the 3.2A/ means that the suit will protect the wearer for a longer time or from a hotter flame. A 3.2A/1 firesuit will take three seconds of 1,800°F heat before the wearer sustains second-degree burns. A firesuit with a rating of 3.2A/5 protects the wearer for 9.5 seconds under the same conditions. Drivers who compete in series using alcohol fuels—which burn much hotter—may wear 3.2A/20 suits, which are rated for 40 seconds of protection at 1,800°F. Regardless of the rating, the material from which the suit is made has to extinguish itself within two seconds after a flame is removed. This means that the firesuit has to substantially decrease or eliminate one (or more) of the three components (fuel, heat, and oxygen) necessary to sustain a fire.

The first fire-retardant suits were regular cloth sprayed with fire-retardant chemicals; however, these chemicals gradually washed off, so the firesuit offered less protection each time it was cleaned. Although fire retardants that stayed on fabrics longer were developed, most firesuits today are made from materials that are inherently fire-resistant.

Nomex is a fire-resistant polymer developed by DuPont in the 1970s. Nomex is a first cousin to Kevlar. The two molecules contain the same atoms; however, one small change in the way the atoms are connected makes a huge difference in their properties.

Kevlar polymers are straight, while Nomex polymers have a kink. The straight molecules align well and form good bonds between molecules, which make Kevlar fibers very strong; however, Kevlar starts to melt at about 900°F (482°C). The bend in Nomex molecules makes them harder to align, so Nomex is not as strong as Kevlar. However, Nomex doesn't melt.

When the temperature reaches about 800°F (427°C), a carbon coating forms around the outside of the Nomex fiber. Carbon is a good thermal insulator and can withstand temperatures higher than 6,000°F (3,316°C) without melting. Heat energy is used in forming the carbon layer, which decreases the heat transferred to the driver. The carbon layer separates the interior of the polymer from the flame, effectively cutting off the fuel supply. Finally, the fiber gets larger as it grows its carbon shell, which closes the spaces between the woven fibers and decreases the amount of oxygen available.

CarbonX, which was developed by Chapman Innovations, is a slightly different type of fire-retardant polymer. CarbonX (scientific name: oxidized polyacrylonitrile) is pre-charred: The polymer is heated to 1,300°F during fabrication, so it comes with an outer carbon layer already there. Exposing these fibers to high temperatures or flame completes the carbonization process to the center of the polymer.

CarbonX protects the driver the same way that Nomex does; however, CarbonX is even more fire-resistant. CarbonX will not ignite or burn, even when exposed to temperatures exceeding 2,600°F for more than two minutes.

CarbonX provides a lot more protection than your average NASCAR driver needs; however, that's not the main reason most firesuits are made of Nomex. While Nomex can be dyed virtually any color, CarbonX is currently limited to black and a few other dark shades—after all, it's essentially already burnt. It'll work for Clint Bowyer (who wears an all-black firesuit), but not for a lot of other drivers.

CarbonX is used increasingly in the long underwear that drivers wear. Long underwear in a 140°F cockpit may seem pointless, but remember that flame doesn't have to get through a firesuit—high heat causes its own problems. Many popular underwear materials, like nylon, spandex, and polyester, melt at relatively low temperatures. Molten plastic can embed itself in the skin, sometimes to the point of needing to be surgically removed. Fire-resistant underwear doesn't burn, plus its thermal protection gives the driver a few extra seconds to get away from the fire. Many drivers combine CarbonX underwear with a Nomex suit of two or three quilted layers. Racing underwear is important enough to be covered by its own SFI specification. (SFI 3.3, if you're curious.)

All of the firesuits—everyone on the team wears them during races—are stored in cabinets inside the hauler because ultraviolet light, including that from the sun, degrades the polymers. The energy from ultraviolet light can be absorbed by the polymers and used to form free radicals. Free radicals, which are particularly chemically reactive atoms or molecules, wander around trying to form alliances with other atoms or molecules. They aren't shy about breaking up the atoms in a polymer like Nomex, which degrades Nomex's fire-resistance.

Of course, the best way to protect drivers from fire is to prevent fire. The gas tank—technically called a fuel cell—is designed to minimize the chances of a driver needing his firesuit. The fuel cell consists of a metal box with a flexible heavy-duty "bladder" that holds the gasoline. Steel tubing from the chassis forms a protective box around the fuel cell. The interior of the bladder is filled with porous foam that prevents the fuel from sloshing around in the tank and slows its release if the bladder is breached. The fuel cell inlet has a flapper like that used in a toilet tank. The flapper closes automatically if the car tips, thus keeping the fuel contained in the tank. The fuel cell for 2007 holds around 17 3/4 gallons of gasoline and has been moved further away from the rear bumper. Energy-absorbing aluminum honeycomb materials have been added around its perimeter. Even the fuel pump is chosen with safety in mind: It is mechanically driven so that if the engine shuts off, fuel stops being pumped.

Despite these precautions, fire sometimes does break out, so each car is equipped with a cockpit-triggered fire extinguisher and an automatic fire extinguisher near the fuel cell. Some cars also have nozzles in the engine compartment. The combustion reaction that drives fire is mediated by highly active hydrogen atoms. Halogens— the elements in the column of the periodic table that include fluorine, chlorine, bromide, iodine, and astatine—are good at tying up the overactive hydrogen atoms, which interrupts the burning reaction and quenches the flame. In-car fire extinguishers used to use chlorofluorocarbons; however, chlorofluorocarbons deplete ozone in the atmosphere. NASCAR fire extinguishers now use a DuPont fluorocarbon that works the same way but is more environmentally friendly.

Ramon pays special attention to the filter through which air passes before it enters Elliott's helmet. The air in the cockpit is hot, so air is brought into the car from the right rear quarter window. A blower that runs off the car's battery pulls the air through an air duct, and after passing through the filter, the air travels through a hose directly into the top of Elliott's helmet.

Carbon monoxide (CO) is produced by incomplete combustion of gasoline. Combustion should turn hydrocarbons and oxygen into carbon dioxide and water, but it only works when the numbers of hydrogen, carbon, and oxygen molecules are perfectly matched. If there is too much or too little oxygen, or the gas doesn't combust fast enough or completely enough, an engine can produce carbon monoxide (CO), volatile organic compounds (VOCs), and nitrogen oxides. About 2 percent of the average car exhaust is CO. VOCs and nitrogen oxides produce smog, which is a long-term health issue. CO poses a much more immediate threat.

Hemoglobin is your body's taxi for oxygen molecules. Hemoglobin moves oxygen from the lungs to places like the heart, brain, and muscles. When the hemoglobin reaches its destination, the oxygen hops off and the hemoglobin goes back to the lungs for more oxygen. Carbon monoxide also likes the hemoglobin taxi, but once it gets in, the CO doesn't want to get out again. CO binds to hemoglobin 240 times more strongly than does oxygen, which leaves fewer opportunities for oxygen to get to other parts of your body.

Passenger cars use catalytic converters to convert CO to less harmful gases like carbon dioxide (CO_2). Race cars don't have catalytic converters, so drivers are more likely to be exposed to CO, especially if an accident pokes holes in the car's body and crush panels.

CO, like most poisons, can affect you acutely or chronically. Acute exposure is a high dose over a short time, while chronic exposure is multiple small doses over long periods of time. In chronic CO poisoning, each successive exposure further impairs the person's ability to distribute oxygen throughout the body.

Carbon monoxide is tasteless, odorless, and invisible, so how do you know when your driver is being exposed? Ray Evernham, when he was crew chief for Jeff Gordon, could tell immediately, "by the way Jeff answers me on the radio, when the carbon monoxide is getting to him. He becomes a smart-ass. When I started working with him, I thought he was a smart-ass. But the more I got to know him, and the more I learned about carbon monoxide, the more I realized what was happening." Crew chiefs may notice a driver delaying answering questions, ignoring instructions, or making irrational decisions (although some drivers behave that way normally).

Breathing pure oxygen, which increases the number of oxygen molecules in the lungs, is the primary treatment for CO exposure. More severe cases sometimes use hyperbaric oxygen therapy. The driver is placed in a sealed pressure chamber, similar to those used for treating decompression sickness in divers, and exposed to high-pressure oxygen. Advocates claim greater penetration of oxygen at tissue level, which displaces the carbon monoxide–laden hemoglobin from the red blood cells more quickly; however, there is disagreement in the medical community about the efficacy of this treatment.

Many drivers breathe oxygen before and after races, especially at short tracks. Jimmie Johnson, for example, breathes concentrated oxygen for an hour each day he's at the track, and sometimes after the race. We're still learning about how CO affects the human body, so we don't know for sure whether (or how much) this helps. As Jimmie says, "Everybody has different beliefs with it, and it hasn't scienced out to a T yet."

CO exposure is determined by measuring the amount of CO a person exhales, or by measuring the ratio of carbon to hemoglobin molecules in the blood. A 5 percent concentration of carboxyhemoglobin (carbon monoxide bound with hemoglobin) is enough to decrease motor skills and exacerbate heat stress.

Adding an oxygen atom to carbon monoxide (CO) changes it to carbon dioxide (CO_2). Unfortunately, oxygen comes in molecules (O_2), and the oxygen atoms are more than happy to stick with each other rather than joining up with the CO.

A catalyst is a material that makes a chemical reaction occur (or helps it occur more quickly) but doesn't take part in the reaction itself. A catalyst is like a matchmaker: It facilitates the reaction, but doesn't participate in it. Platinum catalyzes the carbon monoxide/carbon dioxide transition. When oxygen molecules encounter a platinum surface, the two oxygen atoms let go of each other and hang out separately on the platinum. A CO molecule coming by can grab one of those oxygen atoms. The oxygen atom would rather join the CO molecule than be stuck on the platinum surface, and the CO becomes CO_2. Since the reaction occurs only at the platinum surface, catalysts are made in very small pieces or coated on a honeycomb support to maximize the available surface area.

Race cars don't have catalytic converters for the same reasons they don't have mufflers: A catalytic converter would slow the rate at which exhaust gases could be cleared from the engine. Instead, the driver gets his own catalytic converter, but it's a little different than the converter on a car. Platinum doesn't catalyze the CO to CO_2 reaction until it gets to temperatures around 550°F to 750°F. At lower temperatures, CO sticks to the platinum. At higher temperature, the surface is too hot for the CO to stick, but hot enough for the oxygen molecules to split up and stick.

The solution came from outer space. Not literally, but from a NASA project to design a space-based carbon-dioxide laser for atmospheric monitoring. Carbon-dioxide lasers convert carbon dioxide to carbon monoxide. Since it's tough to arrange regular deliveries of carbon dioxide gas to space, the scientists needed a way to convert CO back to CO_2 so the laser would have a continuous CO_2 supply.

Space is very cold, so NASA scientists had to develop a catalyst that worked at lower temperatures. While they were developing this catalyst—as often happens—someone else invented the solid-state laser, which is cheaper, smaller, and doesn't need carbon dioxide. NASA adopted the solid-state laser and didn't need a low-temperature CO-converting catalyst anymore, but it was the perfect solution for NASCAR.

Joe Gibbs Racing and Penske Racing helped develop two of the CO-removing systems for race cars now on the market. The manufacturers estimate that using these systems filters out 70 percent or more of the carbon monoxide. In 2006, then–managing director of competition for NASCAR Gary Nelson reported that the level of carboxyhemoglobin in drivers' blood averaged about 3 percent. The average level for a nonsmoker should be less than 2 percent.

CO isn't just a problem on the race track. Anything that uses combustion, including gas furnaces, kerosene heaters, generators, and even burning wood in your fireplace can produce CO. High CO concentrations can cause death, which is why you should make sure the CO detectors in your house work.

Elliott has to check his own CO detector at home, but at the track, he's in extremely capable hands. Ramon attached the air hose to the top of Elliott's helmet and checked that the connection was secure. He hung the helmet on a hook inside the cockpit and the crew was ready for the last hurdle before the race: tech inspection.

Seventeen

The Locusts in the Pit Box

"They call them the locusts," Kiwi, the shock specialist, told me with just a hint of a grin, "because they arrive on Sunday, eat everything, and then leave."

He was referring to the pit crew—the guys who go "over the wall" to change tires, fuel the car, and make adjustments. Three members of the garage crew also serve on the pit crew. Chris is the gas man and works with the catch-can man, Todd Anderson, who does setup on the No. 19 car during the week. Tony and Kiwi change the rear tires while Terry Spalding and Brett Morrell do the same in the front. "Big Ed" Watkins is the jackman.

The morning before, Josh, the team director, had instructed me in the fine art of pit-box selection strategy. Josh explained that crew chiefs choose their pit boxes in the order their cars qualify. The No. 19 team had qualified 18th, which made planning pit strategy more challenging than when they qualified second in Atlanta.

Josh wasn't the only one out for a morning stroll the day before. Chad Knaus, Jimmie Johnson's crew chief, was pacing off boxes to check for size differences. Josh was more interested in location. He had known at Atlanta that he'd get an opening—a pit box with an empty space either in front of it or behind it. Openings make it easier for the driver to get in or out of the pit box, but at Martinsville, boxes with openings will be gone by the time he gets to choose.

There are three important lines on pit road. Josh pointed them out with his coffee cup: One line is the imaginary extension of the start-finish line on the track across pit road. The driver gets credit for completing a lap as he crosses the start-finish line. If he's on pit road, this line is where that happens. The commitment line is at the entrance to pit road, and the exit line is at the end. When pitting under caution, the order in which you cross the exit line determines your position for the restart, so getting out of your pit box and over that line quickly is important.

The pole-sitter almost always picks the first pit box—the one closest to the exit—at Martinsville. The exit line is only a few feet away from the first pit box, so the car can pick up a few positions just by getting its nose over the line a few inches before anyone else. At Atlanta, the exit line is twenty feet away, and having a box a little farther back can be advantageous. You can get up to pit-road speed before the guy in the first box can get going, and you might beat him across the line. Josh prefers pit boxes past the start/finish line. If the cars ahead of you have pit boxes before the line, he explained, you can get the extra five points for leading a lap because you'll cross the start-finish line just getting to your box.

Once he narrowed down his choices, Josh checked each candidate pit box for bumps or missing chunks of concrete that might trip up a crew member. What you'd really like, Josh said, is to have a car you know is going to be off the lead lap (meaning it's been passed once or more by the leader) pitting in the box behind you. Lead-lap cars get to pit first, which means the box behind you will be empty and getting in will be easier. This strategy, of course, assumes you'll be on the lead lap throughout the race.

Once Josh has picked a box, the pit crew's primary task on Sunday is setting up the pit box. Although the garage crew has been here since Thursday, the pit crew has just arrived and pit road is the

Sunday morning social hotspot. Big Ed, a persistently smiling blond Virginian who was an offensive lineman at East Carolina University, waves and shouts welcomes in a booming voice to crew members from other teams.

The "war wagon," which will be the on-track command center for the race, started the morning as an unremarkable box about six feet tall, twelve feet long, and four feet wide. The crew unfolded the box, and in minutes, it became about four times its original size. The top of the wagon, where Josh, Kirk, and Chad will sit, is accessed by ladder. The front bench seats four, with monitors for each person. Back benches in the corners seat two more people, and the entire second story is covered by a canopy. The front of the war wagon, which faces pit road, looks like a tool chest. It has slide-out drawers, the gas cylinders for the air wrenches, and a video player/recorder and monitor. The video recorder is connected to the camera that the pit crew has mounted on a pole hanging above the pit stall. Every pit stop will be recorded.

The back has more drawers for tools, and a plasma television, which would shortly be connected to a satellite dish. They also have Internet access, which makes the pit box about as well equipped as a college dorm room. (It's not that much smaller, either.) The left rear of the box has a tire hub mounted near the bottom for the tire-changers to practice on. War wagons start at about $50,000 and may carry another $10,000 to $40,000 in electronics.

Quinton Mullins, who helps out in the pit box wherever needed, and Swifty, the tire specialist, were bringing the race tires to the box so that the pit crew could start cleaning the wheels. A big, cone-shaped piece of sandpaper on a cordless drill is used to clean the large opening that slips over the wheel hub. The sandpaper is replaced by a wire brush to clean each of the five smaller holes that fit over the lugs. A graphite lubricant is then sprayed on the back side of the wheel around the holes.

After cleaning any stray lubricant from the front of the wheels, the pit-crew members sat amidst the tires with a large box of lug nuts and started attaching lug nuts to the wheels with weatherstripping glue. The glue holds just well enough to keep the lug nuts on the wheels, but doesn't provide too much resistance to tightening with the air wrench. I counted nine sets of tires, so I knew they would be there awhile and headed back toward the garage.

The garage is less frantic on race day than on qualifying and practice days. The most important task for the garage team (Kiwi, Tony, and Chris don't officially become over-the-wall guys until the race) is finalizing the setup and getting the car through inspection. Chris was getting ribbed about his bowling score from last night's team outing. "It was so low," Kirk said in response to a dispute about the value, "that the actual number doesn't matter." In between fending off attacks on his bowling skills, Chris told me that Billy Pink, the company's head engine-tuner, had sent Tom Engleson, the No. 19's engine-tuner, around the garage yesterday to borrow a "valve-spring rubber"—which even I know doesn't exist. I asked Tom, who looks more like an earnest law student than an engine-tuner when he's wearing his glasses, about it. He laughed and clarified the situation.

"It wasn't a valve-spring rubber," he said, smiling and looking just a little sheepish, "it was a rubber spring shim." That also doesn't exist, but it sounds a little more plausible—especially coming from your boss.

The car had to be in line for inspection by noon. After some gentle urging from the NASCAR officials, the team pushed the car out of the garage at 11:55 A.M. As soon as they wheeled the car into line, the rear shocks Kiwi had turned in to NASCAR the previous night were returned. NASCAR requires each team to submit their rear shocks for inspection prior to the race. Kiwi reinstalled the shocks while the car waited in line.

The prerace inspection is similar to the prequalifying inspection we had gone through Friday. There are about twelve stops during which NASCAR officials check to make sure the sticker that was placed over the fuel cell on Friday is still intact, the fuel line length is measured and the ground clearance, wheelbase, weight, and height are checked.

Most of the garage crew followed the car through inspection. Kirk carried a toolbox in one hand and a fluorescent-orange rubber mallet in the other. I thought the mallet might be for chassis adjustments or simply for intimidation, but I soon learned its true function. The hammer goes under one of the car's wheels to keep the car from rolling when we stopped, which we did frequently. Going through tech line is like waiting to get out of the parking lot after the race. You move a little, and then you sit. You sit a lot more than you move.

The primary template grid that fits over the car isn't checked in this inspection because it has to be fitted before the rear wing is mounted. The teams receive the rear wing in the tech line after satisfying this template during prequalifying inspection. The officials do recheck the nose and the rear-end grids.

It took about an hour and twenty minutes to get through the inspection line. The crew cleaned the car between stops more out of boredom than because it was dirty. As the car rolled past the last stop and headed toward pit road, Tom pulled up alongside with a generator—that's the large box you see sitting near the cars on the starting grid. The pit-road generator heats the car's fluids to the appropriate temperatures before the driver starts the engine. The generator cord plugs into a cable in the car that runs into the oil tank: Probes on the inside of the tank heat the oil and thermocouples measure its temperature. It takes ten to twenty minutes to get the oil up to operating temperature.

At 2:00 P.M., all of the team members—dressed in red Dodge firesuits identical except for the names on the backs—lined up across pit road for the invocation. The national anthem followed. Everyone stood, even before the announcer asked them to. That the national anthem was played on the saxophone by a local middle-school teacher instead of "interpreted" by a pop star tells you something about the character of Martinsville.

The end of the anthem featured the traditional flyover by two F-16s. The planes are never where you expect them to be when you look up because light waves travel about a million times faster than sound waves. The light reflected by the plane reaches your eyes almost immediately, while the sound takes longer to reach your ears.

Elliott had a last-minute consult with Kirk and then got in the car. Ramon put up the window net. Richard Petty graced the field with those "most famous words in motorsports"—"Gentlemen, start your engines"—and the parade laps began.

The most important part of the parade laps is establishing pit-road speed for the drivers. Race cars don't have speedometers. A tachometer, which displays the engine rotation rate, is much more valuable to the driver. Pit-road speeds, however, are given in miles per hour and range from 30 mph on lower-speed and smaller tracks to 55 mph on superspeedways.

The field split into two groups, with a second pace car leading the cars in the back half of the field so that everyone was moving at approximately the same speed. The pace car radios the NASCAR tower when it is at pit-road speed, and they relay that information to the spotters.

"Pit-road speed . . . now," Brett, Elliott's spotter, said.

"First gear, five thousand." Elliott responded, meaning that he's in first gear and his tachometer reads 5,000 rpm.

"That'll probably be three thousand nine hundred in second gear,"

Josh said after a moment. They would come down pit road in second gear for green-flag stops and first gear for yellow-flag stops, where they don't have to slow down so much. Josh explained to me after the race that staying in second during green-flag stops minimizes down-shifting and lessens the chance of transmission problems.

The team stood on the pit wall and waved when Elliott passed the pit box to help him identify where they were located on pit road. Drivers have been known to drive right past their pit boxes. Brett and Josh decided that Brett would hand off spotting duties to Josh on pit road when Elliott was ten pit boxes away.

The sun poked out for the first time, but threatening clouds loomed not far off. At 2:17 P.M., Brett said the words that really begin the race.

"Green. Green, green, green."

The yellow flag waved on lap 2 when Casey Mears spun, but it was too early to pit. There wasn't another opportunity until lap 47, when Bobby Labonte scraped the right side of his car against the turn-1 wall.

"I'm sort of thinking of staying out," Josh said. "There will definitely be more cautions before we need to pit." Staying out let us lead the race on the restart (and get five bonus points), but the lead lasted only three laps before we yielded to cars on fresher tires. The race continued uneventfully. The pit crew looked bored, but they perked up at lap 90, when Robby Gordon got a flat tire, the yellow flag waved, and—finally—it was time for a pit stop.

The pit crew used to be whoever from the shop could travel that weekend. The Wood Brothers realized in the 1960s that getting in and out of the pits faster could gain you valuable positions on the track, and they started orchestrating pit stops. Ray Evernham took the concept further as Jeff Gordon's crew chief by hiring pit crews that didn't work in the shop during the week, hiring a coach, and requiring regular practice. Many pit-crew members today are former college athletes hired specifically for their hand-eye coordination and agility.

"Closed. Closed. Closed." Brett repeated. Elliott can't come in until pit road opens. Brett often says things more than once because it can be hard to hear over the radio.

"Swifty, I want a half-pound out of the right rear," Josh said, then repeated, "Half-pound. Right rear."

What Josh meant is that he wanted the tire pressure in the right rear reduced by half a psi (pound per square inch). Pound is just shorter. The pits opened on the next lap and Brett reminded Elliott to watch his speed.

"Thirty-nine hundred, second gear."

Before the 2005 season, NASCAR officials enforced the pit-road speed limit by randomly checking cars using stopwatches. They now use RFID—Radio Frequency Identification—to catch speeders. The system uses the same technology that keeps track of cars during the race. "Scoring loops" are loops of wire buried in the track. Each rectangular loop runs the width of the track, with the long sides about eight to ten inches apart. The loops are spaced at intervals varying from an eighth-mile (660 feet) to a quarter-mile (1,320 feet) and are buried about twelve inches below the track surface.

Each car carries a transponder a little larger than a pack of cigarettes that generates its own unique frequency. An alternating electrical current in the scoring loop creates a magnetic field that changes with time. The magnetic field from the loop induces an alternating current in the transponder, which is the transponder's cue to send its information. That information is picked up by the scoring loop and sent to the master computer. The accuracy of this scoring system, which works at speeds up to 310 mph, is three ten-thousandths of a second.

Similar RFID technology is used in the microchips that identify your pets if they stray from home. Any vet or animal shelter can scan the transponder injected into the dog or cat and find out where the wayward pet belongs. Goodyear uses the same technology to track their tires, as does NASCAR to certify car chassis.

The scoring loops measure the car's average speed, which is the distance between two loops divided by the time it takes the car to travel between those two loops. If you drove the thirty miles from Charlotte to Mooresville in thirty minutes, your average speed would be 60 mph; however, you probably didn't travel the entire distance at that speed. You may have been going 70 mph on the expressway, but slowed down when you got off the expressway, giving you an average speed of 60 mph.

The pit-road speed limit is a safety precaution: There are a lot of people on pit road during stops and a lot of cars swerving around each other. The pit-road speed limit at Martinsville is 35 mph, but NASCAR allows a 5 mph cushion, so a car's average speed between each pair of loops has to be less than 40 mph. You might get away with going 45 mph at the start of a segment if you slowed to 30 mph toward the end of the segment. If you're not careful enough and get caught speeding, you have to pass through pit road at pit-road speed while the other cars continue around the track at their normal speeds.

Before scoring loops, cars restarted after an accident in the order in which they got back to the start-finish line, which was called "racing to the yellow." Scoring loops allow NASCAR to "freeze the field" following a caution. The cars are ordered according to the last scoring loop they passed on the track when the caution was called. Not racing back to the yellow allows emergency personnel clear access to the track.

"Okay, Josh, he's all yours," Brett said when Elliott got ten pit boxes away.

"I got him," Josh said. "Okay, Elliott, five . . . four . . . three . . . two . . . one. Wheels straight, put on the brake."

The tire-changers can't get the tires on and off if the wheels aren't straight. Elliott, like most drivers, has a piece of tape on his steering wheel. When the tape is straight up, he knows the wheels are straight. He keeps the brakes on, even though the tires are off the ground, to prevent the wheels from turning while the tires are being changed.

Seven people are allowed over the wall at any one time. NASCAR sometimes allows teams an "eighth man," who can clean the grille, give the driver a drink, or pull a tear-off from the windshield. Kirk usually serves as the eighth man, although Ramon confided that he often pulls more than one tear-off from the windshield at a time.

All of the windows on the car are made of a polymer called LEXAN, which was discovered by GE scientist Daniel Fox in 1953. LEXAN is 250 times more impact-resistant than glass and about half the weight of the same size piece of glass. LEXAN is used for airplane windows, lenses for glasses, CDs and DVDs, and iPods. Since it is more impact resistant, you can use a thinner piece of LEXAN than you would glass. The front windshield is about a quarter-inch thick and weighs about 11 pounds. The rear and side windows are slightly thinner. Metal bracing is placed inside the windows to prevent them from flexing during the race.

You can fire a .44-caliber handgun at point-blank range at an inch-thick sheet of LEXAN without damaging it, but you can scratch the same piece of plastic with your fingernail. The thing that makes LEXAN impact-resistant also makes it easy to scratch. It doesn't shatter when hit because the bonds between atoms are very strong, but they are also flexible. Instead of breaking, the bonds stretch or compress, which allow the atoms to move. Just as flexible trees can withstand strong winds because they bend instead of breaking, the polymer chains in LEXAN dissipate the energy from a flying piece of debris by rearranging themselves. An impact that would shatter a glass window merely leaves a mark on a LEXAN window.

There is so much dust, sand, rubber, and other debris on a race track that the windshield is almost constantly being pelted by small objects. One of Ramon's jobs at the shop is applying tear-offs—thin sheets of clear plastic that protect the LEXAN windshield. Multiple layers are used because pit stops have gotten so fast that there isn't time to clean the windshield. Removing a tear-off gives the driver a clean view. The fluorescent triangles flapping on the top or bottom of a windshield that you see through the in-car camera are tabs placed on the tear-offs to help the pit crew quickly find the edges. Tear-offs are being used by the United States Army for Blackhawk helicopters. A new helicopter windshield costs between $3,000 and $7,000, while a tear-off costs about a hundred dollars and can last four to six months.

Tear-offs were introduced in 1997 by a company called Pro-Tint, which is also developing the Army application. Tear-offs are made from a DuPont polymer called Mylar. Mylar is a trade name: The polymer's scientific name is "biaxially oriented polyethylene terephthalate polyester." Polyethylene terephthalate (aka PET) is commonly used in food containers. "Bi" means two, so biaxially oriented means oriented along two axes. The PET polymers, which start out like tangled spaghetti, are simultaneously heated and stretched in perpendicular directions to make Mylar. This process aligns the polymer chains in two directions in the plane of the film, making a material that is thin enough to see through but strong enough to remove without tearing during a pit stop.

Tear-offs are usually between 2 mils and 4 mils thick. The number of tear-offs that can be layered is limited by the need to see through the stack. About 4 percent of the light reflects from the surface of the tear-off. That means that 4 percent of the light reflects when it goes from the air to the tear-off and another 4 percent reflects coming out of the tear-off. You lose about 8 percent of the light for each tear-off because there is an air gap between the layers. One way around this problem is to tightly laminate the tear-offs together to displace any air between the sheets. Ramon may apply between four and ten layers of laminated tear-offs, depending on their thicknesses and how much air has been removed between the layers.

The pit crew was to change four tires, fill the gas tank, and add two rounds of wedge to the right rear during this stop. The front and rear tire-changers started removing the right-side lug nuts before the tires were even off the ground. The front and rear tire-carriers were standing by with the new tires, oriented to match the five holes with the five lugs on the wheel.

Big Ed jacked up the right-hand side of the car with a single pump of the jack. A pit-road jack is lighter than a garage jack, with an arm about a foot longer for extra leverage. The tire-changer removed the old wheel and rolled it to the side. The tire-carrier planted the new wheel on the lugs and then directed the old wheel back to the pits where another crew member grabbed it. The tire-changer tightened the lug nuts, which are long and don't have any threads for the first ¾" to prevent the lug nuts from getting cross-threaded. A cone placed over the hub helps center the wheel.

Chris, the gas man, placed a wrench in the jack-bolt hole in the right rear window. Once he sent the old tire rolling toward the pit wall, Kiwi reached up and cranked the wrench two full turns. As soon as the tire-carriers indicated that both tires were on, Big Ed let down the car and the tire-carriers and changers ran to the left-hand side of the car.

NASCAR requires lug nuts to be yellow so that the NASCAR

official in each pit can see that five lug nuts get on each wheel. A number of teams use colored lugs or draw lines from the center of the hub to each lug. They say it helps the pit crew spot their targets faster. Although different teams use different techniques, they all use fluorescent colors—usually pink.

Fluorescence is the process by which an atom absorbs ultraviolet light and emits visible light. Instead of just reflecting the incident light like most molecules, fluorescent molecules absorb light over a range of wavelengths and emit light over a more concentrated band of wavelengths, which is why fluorescent objects look so bright.

If you've noticed that yellow traffic signs look a little different these days, that's because research has shown that black on fluorescent yellow-green is much more attention-getting than the banana-yellow color they used to use. Black on fluorescent pink is the second-most attention-getting combination. The government has adopted fluorescent pink and black signs as the new standard for warning drivers of road emergencies. Fluorescent pink is the preferred color of NASCAR teams as well.

While the tire-changers were working, Chris had brought the first gas can over the wall and inserted it into the gas inlet on the left-hand side of the car. The gas cans have a dry break at the end, which prevents fuel from flowing unless the can is pressed against the inlet. Todd, the catch-can man, inserted the catch can in the rear of the car and held the first gas can while Chris went back for the second one.

When a car arrives on pit road with an empty fuel cell, it really isn't empty—it's filled with air. You can't fill the fuel cell with gasoline unless you give the air a way out. The overflow line from the fuel cell goes to a tube into which the catch can is inserted. The gas cans have a dry-break that allows air to escape.

The handle on the pump at your local gas station also has to allow air to exit the tank—in fact, that's how the gas pump knows to shut off. There is a small hole in the outer pipe. When the hole is covered by the gasoline, that's the signal that the tank is full. Todd serves that purpose for the No. 19 car. When gas started flowing into the catch can, Todd shook the empty gas can up and down to let the jackman and the crew chief know that the fuel cell was full.

"Go. Go. Go. 3,900, 3,900, 3,900. One line, one line, one line." Josh intoned, as Elliott squealed out of the pit box in a puff of smoke. The entire process took about fourteen and a half seconds. Swifty and Ramon scraped the tires and measured the wear indicators. I could feel the heat radiating from the tires—and I was two feet away. The wear numbers were passed up to Chad, Kirk, and Josh.

The extra fuel from the catch can was returned to the gas can and the gas can was weighed. Subtracting the weight of what's left from the weight of the can before the pit stop lets Chad calculate how much fuel made it into the car, and the car's gas mileage. Sunoco provides the team with the fuel's density on the morning of the race to facilitate these calculations.

The over-the-wall guys watched the pit stop on the video player with their pit coach, Brian Feree. There had been a minor problem getting one of the front tires back into the pit box, and they reviewed the tape to make sure they don't repeat it on the next stop.

Between stops, the pit crew rested, watched the television to see how Elliott was running (you can't really see much of the race from the pits), and ate. Fruit cups and pretzels were popular, along with lots of water. Tony practiced on the hub attached to the war wagon. Kiwi stretched. Allen—who helps manage the tires and hoses from behind the wall—was restless. The garage crew had been here since Thursday.

Josh, Chad, and Kirk didn't have time to be bored. The car was misbehaving and Elliott was getting increasingly frustrated.

"We really pissed the car off," Elliott said on lap 156. "We took every bit of forward bite out of it," meaning that the car was not turning. Kiwi sighed and looked down. Allen pursed his lips. Swifty shook his head between tire measurements. Drivers never see the faces of their pit crews during a race, so they can't appreciate how much their comments affect their crew. The leader caught up with us on lap 197 and we watched the car being lapped on the television.

Elliott's mood was a lap down as well. On lap 217 he said, "We keep making it worse all the time." Josh proposed some "big adjustments."

"It doesn't matter if we can't get our lap back," Elliot radioed dourly.

Between the dark clouds looming over the track and the petulant car, the crew morale was getting more and more pessimistic. When a caution came out on lap 305, shortly after a big adjustment, Josh asked, "I'm assuming we didn't help it at all, right?"

"I'm willing to try anything," Elliott said firmly, "I wish we were making it better just once."

"I know you're doing the best you can. I wish we had a better car for you."

"Nothing seems like it correlates over," Elliott sighed. "Changes we made in Lakeland don't transfer to Martinsville."

Josh is right: Setting up a car is just not that simple. It's not unusual for something that worked well during testing to not work at all during a race because the conditions are ever-so-slightly different.

At this point, I was thinking that what Elliott and Josh needed most was a time-out. Four laps later, the skies opened up and the race was red-flagged. The cars came down pit road and NASCAR allowed two crew members to come out and cover the car. The rest of the crew scrambled to set up a canopy over the pit box and get plastic over the computers.

The downpour stopped as suddenly as it started. I thought the rain might lighten the traffic out of the track, but the seats refilled quickly. The crowd cheered the jet dryers when they came out to dry the track. The pit crew inhaled an entire bag of chips in less than eight minutes.

The race restarted, but the most interesting thing on our part of pit road was watching the No. 96 crew's expertise with the most delicate of metal-shaping tools—the baseball bat—after their car was unceremoniously dumped by Juan Pablo Montoya. The crew member banging dents out of the left front fender at least seemed to be getting some satisfaction from his "adjustment."

With 27 laps to go, we were running 24th.

"I think it's the best it's been all day," Elliott radioed. Josh looked over the lap times and responded, "We're the best car on the track right now."

If the race were another couple hundred laps, we might have had a chance at getting back to the front, but we were rapidly running out of time. We had enough fuel to make it to the end of the race, and unless there was an accident, there wouldn't be another pit stop. The crew started packing up. After he finished his duties, Tony stood on the stairs of the pit box and intently watched what Kirk, Chad, and Josh were doing.

If a career in NASCAR is a mountain, there are two types of climbers. One group is satisfied with how high they've climbed and is ready to head home. They're happy with their current jobs, but they are ready to settle down and spend more time with their families. They'll stay on the road another couple of years and save up enough money to start a business of their own, or take a shop-based job.

Tony is in the second group, which is the group that's still headed up. He is smart, has an inquisitive nature, and asks a lot of good questions. I was not at all surprised when Kirk told me that Tony will make a good car chief. I think you'll hear more about Tony in the future.

Jimmie Johnson edged out Jeff Gordon for the win, but we didn't even stay for the burnout. The garage crew was on their way back to the hauler before Elliott had crossed the finish line. The pit crew cleaned up the pit box and packed up the war wagon. Elliott drove the car onto the hauler lift and the crew descended on it. There was a donut—a black circle where another car's wheel had rubbed up against the door—on the driver's side, but the car was otherwise intact. The transponder and the wing had to be returned to NAS-CAR and the tires to Goodyear before the hauler would be allowed to leave. Elliott got out of the car, leaving his helmet and gloves inside, and headed off to his motor home.

Elliott wasn't the only one anxious to get moving. I had never seen guys change clothes so quickly before, but they had extra motivation. The upcoming Easter weekend was one of the three week-ends the crew had off during the season. Some of them would be on the road, but it would be to catch up with family and friends. After that, it would be time to head to Texas and fifteen straight weeks of racing before they get another weekend off.

Eighteen
The Checkered Flag

You've probably noticed that I haven't said much about how the No. 19 team finished. As they were loading the car on the hauler at the end of the Atlanta race, an older gentleman from the No. 21 car next to us asked me where we finished.

"Eighteenth, I think." I must have looked disappointed. He sighed.

"Well, I guess that's better than thirty-seventh."

We finished 24th in Martinsville, one lap down. Chris, the hauler driver, apologized for not giving me a better finish to put in the book. It didn't bother me: After all, the science is the same regardless of where you finish.

By the May 2007 Coca-Cola 600 at Lowe's Motor Speedway in Charlotte, the No. 19 had fallen to 21st place in owner's points. I hadn't planned on watching the race in person, even though I was in the garage that weekend. I planned to show up on Thursday for qualifying and Saturday for practice. Sunday was going to be especially busy: The No. 19 was running a special paint scheme promoting the *Fantastic Four: Return of the Silver Surfer* movie and some of the stars would be in the pit box. It was going to be over 90°F, so I'd be happy watching the race from my air-conditioned hotel room.

The team, however, had other plans for me. They qualified third.

They had the second- and fourth-fastest times in Saturday's practices. Saturday night, Kasey Kahne, Elliott's teammate, won the NAS-CAR Nationwide Series race. The team was more optimistic than I had seen them before. No one dared venture more than, "I think we have a sixth-place car," or "I think we have a shot at it," but I'm sure they all entertained the same thought I did: *This could be the race where we turn things around.*

So despite an impending book deadline and a strong aversion to fighting 200,000 people getting out of the parking lot, I arrived at the track Sunday at 10 A.M. The race didn't start until 5:40 P.M., but if I came much later, I would have spent most of the day stuck in traffic. Josh had advised me where to park to get out quickly. It's not a secret: I was surrounded by crew members in pickup trucks listening to the radio or napping until the garage opened. At about 10:45 A.M., I joined the stream of crew members from the parking lot. It was already steamy: I was dripping with sweat by the time I reached the garage.

I wasn't the only one feeling the heat. Swifty, the tire specialist, was running ragged. I counted seventeen sets of tires sitting in the pit box—and behind the pit box, because they didn't all fit in the pit box. Some were stickers, meaning that they hadn't been used and the Goodyear stickers were still on them, and some were scuffs that Elliott ran for just a couple of laps during practice.

By 5:30, the movie stars and their entourage had filled every space in the pit box that wasn't occupied by tires. The car had a special paint scheme with the Silver Surfer on the hood instead of its normal red-and-white colors. I settled in about twenty feet behind the pit box, nestled among the last row of tires. I put on some more sunblock but realized that even SPF 45 wasn't going to prevent a white mark where my watch was.

Finally, the command to start the engines was given and the race began. We started third, but fell back to 4th at lap 11, 6th at lap 17, and 11th at lap 31.

"Be patient with the car. It'll come to you," Brett, the spotter, said over the radio. The Coca-Cola 600 is the longest race of the year: 400 laps around the 1.5-mile track. I pondered whether dropping back was a strategic move to clear the pit box because the crew could barely move with all the people and tires there.

"*Way* too tight in the center," Elliott said.

"Late center, early exit off two is where we're losin' it," Brett added.

On lap 46, Greg Biffle cut a tire, bringing out a caution. Josh and Elliott debated the merits of a spring-rubber change versus a wedge adjustment. Elliott brought the car into the pit, the pit crew jumped over the wall and completed the service, and then Elliott sped away. The stars and their entourage left, I suspect for the air-conditioned comfort of suites.

The pit stop was slow. Josh apologized to Elliott and promised better next time.

"Three hundred and forty-nine laps to go," Brett reiterated, "so keep takin' care of it."

The team knows from practice that the car wouldn't handle as well during the daytime part of the race. The "Fantasticar," as it had been dubbed, would come to life when the sun started going down and the track cooled.

"This is a long race," Josh said. "A long, long race." Josh, Chad, and Kirk talked over possible changes. The pit crew prepared for the next stop and a chance to redeem themselves. On lap 52, we're in 23rd position—but remember, the crew kept telling each other, it's a long race.

But not for us.

On lap 53, A.J. Allmendinger got into Jimmie Johnson, who was running 5th at the time. It didn't look like much, but then Jimmie started frantically waving out the driver's-side window and slowed down. The cars immediately behind him started going around, but before the information could filter back into the field, the tread from the cut left-rear tire came completely off.

Smoke. Swerving. Squealing of tires. The dull thud of cars hitting the wall and the slightly higher-pitched, stomach-turning crunch of metal hitting metal.

A dozen cars were involved in the wreck and one of them was ours. Elliott hit the outside wall, was pushed into Sterling Marlin, and then rolled down the track where Dave Blaney's No. 22 hit his left front.

If you saw the race on TV, that was Josh banging angrily on the pit wagon. You didn't need to be able to read lips: The ferocity with which he hit the desk told you the level of his frustration. Chad and I had been talking earlier about how hard it is to leave your finish at the track. The look on Chad's face as he started down the ladder told me that he would have a long week ahead of him.

"We're done," Elliott said, his voice heavy with disappointment. Brett—ever the optimist—radioed that the car didn't look so bad; there was damage on the left front, right front, and rear. Apparently the sides of the car were OK. Josh and Brett discussed whether they needed to go to the garage, or if they might be able to fix the car in the pit, but Elliott interrupted decisively.

"We one hundred percent gotta go to the garage. One hundred percent."

Kirk took charge, telling the team to get ready to replace the right-front spindle, lower control arm and tie rods, and to get the spare nose ready. As Elliott approached the garage, he found that he didn't have any brakes. He slowed down as much as possible and the team caught the car as it went past, then pushed it into the garage.

Josh apologized.

"Sorry, Elliott, we should have been ahead of that wreck."

"I should have done more to get you through that," Brett added.

By the time I got back to the garage, everyone from the pit crew and the garage team was intently working on the car. This was the first time any of them actually looked like they worked on cars for a living. The back of Kirk's uniform was greasy and black from being under the car. They sawed off the damaged nose and replaced it with the spare from the regular paint scheme. The left side of the hood was badly crumpled. Chris, Allen, and Chad were working on the back of the car, cutting away the damaged rear panel and replacing it. Josh conferred with Ray Evernham on one side of the garage. Josh's wife, Michelle, stood a few feet from the garage entrance looking in and nervously biting her lip.

Elliott—thankfully uninjured—emerged from the garage and headed off to wait in his motor home, a sign that the repairs were going to take a while. After three weekends, I could get a pretty good read on Elliott's mood just from the slope of his shoulders and the angle of his head. This was the first time the whole weekend I had seen that dejected slump.

Brett, who continued to monitor the track action, told us that there were only thirty cars running—more than 25 percent of the field was out and some of the cars weren't going to make it back. The No. 16 and the No. 66 were already loading up to leave. At this point, why not call it a day and settle for our current 40th position? Because we could move up at least four places by getting back on the track. Every position means points and every point is important.

One hundred and thirty six laps later, we were back. The Fantas-ticar had become the Frankencar, patched with pieces from the normal red-and-white paint scheme and held together with tape. The team was making the best of the situation.

"Let's see what we can learn about these tires," Josh said, since they would use the same tires when they return in October. I expected Elliott to be in a really bad mood because making laps in an injured car can't be much fun. He was surprisingly sanguine.

I wasn't. I felt inexplicably sad. For the team, for Elliott, for the thirteen sets of tires Swifty prepared that would be returned to Goodyear without even getting on the car tonight.

Watching the replay of the crash during yet another caution, déjà vu took me back to the crash I accidentally happened to see on television, the crash that brought me here tonight rooting for an injured car with a best possible finish of 36th. Even though I was captivated by the science, I remember wondering back then why anyone would possibly want to watch cars run in circles. Now I knew the answer.

It's not about the cars.

It's about the people.

For most NASCAR fans, that's the driver, but I had the privilege of getting to know the folks who make everything possible for that driver: The crew chiefs, mechanics, shop foremen, fabricators, engineers, engine designers, pit crew, hauler drivers, and the marketing and public relations people. The more I thought about how many people had put not only their brains, but their skill, their hearts, and their desire into this car, the more my eyes started to well up. I went for a walk because I was damned if I was going to let anyone see me get teary eyed—especially over a bunch of cars running in circles.

On lap 294, we were 138 laps down when out of nowhere, Elliott radioed to the team, "Don't hang your heads. I promise you better days are coming."

One hundred and six laps later, we finished 36th. The crew started to pack up as Elliott crossed the start-finish line.

"We had a good car, man." Josh said to the team over the radio. It was a simple statement of fact. Then, more wistfully, he said again, "We had a good car."

There was silence for a moment before Elliott responded.

"Ten-four."

Epilogue

Daytona 2008: What's Better Than a Happy Ending?

Every now and then, one of my science acquaintances asks me if I'm done with "that NASCAR thing" yet. Sure, the initial excitement of venturing into a race shop or the NASCAR garage has passed. I've answered the questions that initially got me interested in racing, but it seems the more I learn, the more questions I have. I even started a weekly blog (www.stockcarscience.com/blog) where I can explain things like the tapered spacer, the mysteriously detached Las Vegas oil-tank lid, and the challenges of developing tires for the new car. There is always something new to learn because NASCAR is always changing.

The No. 19 Gillett Evernham Motorsports (GEM) team has had its share of changes. Some changes were obvious, like Elliott Sadler in the Best Buy blue firesuit of his new sponsor. Only Tony Lunders and Allen Mincey are still with the No. 10 team. Tony replaced Kirk Almquist as car director. Kirk started the year in the NASCAR Nationwide Series garage as crew chief for GEM's NASCAR Nationwide No. 9 car. Two former members of the No. 19 crew went with him: Chris Miko (the hauler drive and gasman) and Ramon Zambrano (now a shock specialist). Swifty is a tire specialist for David Reutimann at Michael Waltrip Racing, and Tom Engleson is tuning engines for BAM Racing.

Even the nature of NASCAR is changing. Back in the day, "team" and "company" were synonymous: One company ran one car. No more. All of the companies that can have plans to expand to the NASCAR-mandated maximum four teams. Roush Fenway Racing started the current round of mergers and acquisitions in 2007. Even the venerable Petty Enterprises, after much internal deliberation, brought in business partners. Companies merged engine shops and forged partnerships. Alliances are necessary to survive in this increasingly competitive sport. You dare not fall behind, as GEM knows all too well.

The No. 19 finished 2007 in twenty-fifth place. Even Kasey Kahne, who won six races in 2006, struggled, finishing in nineteenth place. In mid-2007, Ray Evernham sold majority ownership of his company to George Gillett, allowing Ray time to step back from trying to do everything and target his expertise. Eric Warren, GEM's Director of Competition, moved to Michael Waltrip Racing. Mark McArdle, a fifty-year-old Wisconsin native with three Indianapolis 500 victories, became GEM's Vice President and Managing Director of Competition. Tommy Wheeler, a 1999 physics graduate of Davidson College, took over the engine shop and became Director of Engineering Services for GEM. By the beginning of the 2008 season, GEM had a new corporate structure—and a new attitude.

The optimism permeating the No. 19 hauler wasn't surprising when I stopped by just before the Daytona 500. Daytona in February is all about potential. Daytona in July is about reality. After a sixth-place finish at the Daytona 500, a bad-luck cloud settled over the No. 19. They were running better than last year, but their finishes didn't reflect that improvement. After seventeen races, they were back in twenty-fifth place.

Rodney Childers, Elliott's new crew chief, is a quiet, self-effacing North Carolina native who grew up immersed in racing. Rodney sometimes drove in one series and then served as crew chief for someone else. His parents wanted him to enroll at the University of North Carolina–Charlotte to get a mechanical engineering degree. Rodney convinced them to let him attend a community college for a couple of semesters first. He hasn't quite gotten around to putting in that transfer application yet.

Rodney has a quiet confidence that the team has done its homework and is ready to handle whatever challenges arise. Even when Elliott rails about how bad the car is, Rodney always answers with a calm "10–4, we'll fix it when you come in." He has an easy rapport with the team's engineer, Kevin Kidd, a jolly Virginia Tech–trained engineer who also has extensive experience behind the wheel and on the pit box. Their personalities are very different, but there is clear respect for each other's expertise.

The shift from cars to companies is evident during practice. Mark and Tommy scrambled up and down hauler ladders, encouraging the crew and transferring information between teams. After the crew hustled to replace the entire rear-end gear at the end of practice, Rodney and Elliott thanked everyone for the hard work. I heard more "thank yous" during that single practice than I did during entire races last year.

Even the driver had changed. Clean shaven and newly engaged, Elliott was decidedly calmer, much more detailed in his feedback about the car, and clearly happier. He told reporter Marty Smith, "I went from being stubborn about my driving style to changing and adapting better. I've had to change my whole way of thinking." While an afternoon thundershower poured down outside, Elliott hung out in the front part of the hauler trading stories with the crew.

In the back of the hauler, Tommy Wheeler had just returned from a meeting about NASCAR's soon-to-be-announced new testing policy. He and Mark McArdle thought aloud through short-term and long-term testing strategies. Crew members chimed in with ideas and concerns. One mechanic confided that the big difference this year is that people are listening and know that they are in turn being listened to. The results speak for themselves: Two former members of the No. 19 crew, Kiwi Duncan and Chad Johnston, have visited Victory Lane three times (so far) with GEM's No. 9 car.

Having the right combination of people at the right time is important, but you can't overlook the progress GEM has made understanding the new car. After an off season of experimentation, many of the engineers I spoke to had a decidedly more optimistic attitude. I saw that optimism when I visited Josh Browne, who had been Elliott's crew chief during part of 2007 and is now chief race engineer at Red Bull Racing. Josh excitedly drew graphs on his office whiteboard to illustrate how much more they understand about the forces on the new car compared to what they understood last season. I saw the same excitement in other engineers at other shops. The interesting thing is that no two engineers had the exact same theories about how to make the cars faster. Josh and his colleagues must be on to something: Red Bull has made major gains in just their second year in NASCAR.

That isn't to say that the engineers aren't actively lobbying NASCAR for changes. The new car has three major challenges relative to the old one: a higher center of gravity (CG), less overall downforce, and limited front-end travel. The higher CG exacerbates load transfer problems. Braking causes some of the car's weight to shift from the rear to the front. Accelerating causes weight to shift from front to rear, and cornering shifts weight from the inside wheels to the outside wheels. A car's grip is proportional to how hard the wheels are pushed into the track, so when you brake, weight shifting from the back wheels to the front wheels means less rear grip and more front grip.

How much weight shifts is proportional to how high the CG is above to the track. Let's assume the weight for a 3600-lb car is equally distributed between the left and right wheels when the car is standing still. If the car has a CG height of 15 inches, turning with 1*g* of acceleration leaves about 920 lbs pushing down on the left-side tires and 2680 lbs on the right-side tires. (I'm assuming, of course, that the car is turning left.) The total weight is the same, but it's distributed differently when the car is turning. If you raise the CG by two-and-a-half inches, you lose an additional 150 lbs of grip from the left side tires. You gain 150 lbs on the right, but you can only go as fast as the tire with the least amount of grip allows. The higher CG in the new car means more weight transfer and less grip.

In addition to less mechanical grip, the new car has less aerodynamic grip. Aerodynamic downforce helps push the tires into the track, increasing grip. Estimates are that the new car has anywhere from one-half to two-thirds the downforce of the old car. Again, less downforce means less grip.

The final challenge is the lack of front-end travel. The old car wasn't subject to any rules about maximum front height. The new car has both a minimum and a maximum front height, which means that the new car can travel only about half the distance the old car could. One of the important roles of the suspension is to position the car in the most aerodynamically advantageous position. You want the nose of the car close to the track to maximize downforce. It's much harder to do that with the new car than it was with the old car. The overall result is a lot of drivers saying, "We need to get the car to turn better."

If there is one adjective that describes the people who work on these cars, it is "innovative." If you can't make the car turn with your existing bag of tricks, it's time to open a new bag. My favorite (so far) is how engineers figured out how to make the car asymmetric despite rules that were created specifically to limit asymmetry.

The old car was asymmetric because the body was offset on the chassis. The left-side fender was much broader than the right-side fender. Asymmetry helps the car turn; however, it also makes for an odd-looking car. Add to that the cost of aerodynamic research, and NASCAR decided to make the new car much more symmetric and to limit what changed teams can make to the body.

"Yaw" describes the difference between the direction the car is moving and the direction it is pointing. A car with no yaw (shown on the left side of the drawing below) travels in the same direction it points. The car on the right side of the drawing is yawed, which means the car is headed straight, but pointing a little to the left.

A car by definition is yawed when it corners because the car is pointing in a different direction than it is moving at every point along a turn. Air pushes on the car differently when the car is at an angle to the oncoming air compared to when it is hitting the air head-on. Yaw lets the air help the car turn. In effect, a yawed car has a head start on turning.

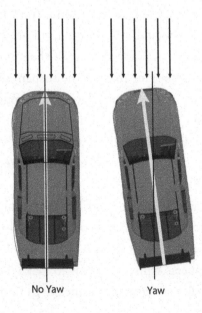

No Yaw   Yaw

Since they can't change the bodies, teams figured out how to modify the rear axle and the angles of the rear tires to make the car "crab" or "dog track." A running dog is a little sideways: The rear paws never follow in the front paw prints. You can do the same thing with a car. If the rear wheels don't follow the front ones, the car will be yawed, even when it is coming down the straightaway. This is accomplished by curving the rear axle and toe-ing the rear tires, which mimics the effects of offsetting the body without violating any of the new-car rules.

Well, almost. If the first adjective that describes NASCAR engineers is "innovative," the last adjective would be "restrained." By Mother's Day in Darlington, some of the cars were so sideways that they couldn't get onto the scales for inspection. NASCAR finally had to make made a rule limiting the amount of yaw teams could get this way.

Yaw wasn't the biggest science story at Daytona: Engines were. Every year around the restrictor-plate races, we hear that some manufacturers are at a disadvantage because their engines "don't make as much horsepower." Which manufacturer is at a disadvantage depends on who is speaking.

The problem with these debates, most of which are of the form "[insert manufacturer] has twenty-five more horsepower than we do" is that they focus incorrectly on one number. Periodically throughout the season, NASCAR impounds a limited number of cars after a race and tests the engines. The number that usually gets reported in the press is the peak horsepower, which is the maximum horsepower an engine makes. The problem is that horsepower is a function of engine speed. The peak horsepower value doesn't really matter if your maximum horsepower is produced at 9,000 rpm, but you spend virtually the entire race around 8,400 rpm because of the gears NASCAR mandates. The issue then is not who has the most peak horsepower, but who has the most horsepower in the rpm range they'll actually be running the race at. That's a piece of information we rarely hear. Elliott was one of the drivers advocating that NASCAR allow teams a wider range of gears to run at plate tracks so that they can get the engine operating in a more favorable rpm range.

Dodge and Toyota took the majority of the top spots at the Daytona 500, which seemed to contradict Elliott's concern about the engines. When I asked Elliott, he argued that finishes at plate tracks don't depend much on engines—they really reflect how well the driver can draft. He told me that the problem is that "other drivers don't want to draft with the Dodges because we're slower." That would imply that Dodge's domination (six of the top ten finishes) of the Daytona 500 was possible only because there were enough Dodges up front that were willing to draft with each other. Unfortunately, it's a little hard to run a control experiment.

Still bothered by this after the garage closed, I consulted with my go-to engine guy, Andy Randolph (who is now engine technical director at Earnhardt Childress Racing Engines). Andy agreed with Elliott and said that the real test of who has the best engine is qualifying; however, it turns out to be hard to make conclusions based on that data, too. The Daytona 500 has a unique qualifying procedure, so we couldn't use those qualifying times. When we looked at the data for the July 2007 Daytona race, seven of the top ten spots (before qualifying was rained out) were occupied by cars that had to race their way into the field.

Daytona is an impound race, which means you can make very limited changes to the car after qualifying. Teams that must have a good qualifying time to make the race resort to things like not hooking up the alternator during qualifying. (They have to make a pit stop to hook up the belt within the first few laps of the race and usually end up in the back of the pack.) Even among the top thirty-five teams guaranteed a spot in the race, some teams are concerned about qualifying well and others aren't, since you can win at Daytona from almost any starting position.

I've come to two realizations since initially writing this book. The first is that I hadn't appreciated just how complex a race car actually is. A race car is a holistic system, like the human body. If your heart isn't working right, it affects your brain and lungs. If you change one thing on a car, even by a small amount, you can affect a dozen other things—some of which you might not have predicted. The systemic nature of the car is one reason, I suspect, NASCAR is reluctant to make changes until the end of the season. Changing the front travel, for example, will require new suspension setups. Those setups may change tire loads significantly, requiring Goodyear to develop new tires. It will be interesting to see how the car changes and which teams are best able to keep up with its evolution.

The other thing I hadn't fully appreciated is the passion and the dedication of the people who work in this sport. Sirius Speedway host Dave Moody asked me (during a conversation about the ever-narrowing "gray area" in which the teams are allowed to innovate) why engineers would keep working so hard if they are continually getting limited in what they can do. The answer is because they absolutely hate losing. It doesn't matter if you weld the chassis, change the tires, do computer simulations or drive the car: Everyone in NASCAR is a competitor.

There is a lot on the line at every race. I was reminded of this watching Andy Randolph pace in the No. 29 pit, three pit boxes down from the No. 19's pit box—before the race had even started. The intense heat had subsided to a tolerable level and the sun sank in the sky. Kevin was on the pit box setting up computers while the rest of the crew lined up on pit road for opening ceremonies. The second and final practice had been rained out, so they never got a chance to fine-tune their setup.

"We'll just use the same setup as February," Rodney said in a quiet drawl when I asked him what they were going to do. Elliott qualified fortieth, but Rodney was unconcerned. "We ran well in February," Rodney said. "We'll do OK."

After some initial issues with trash on the grille, Elliott made his way toward the front. Rodney buoyed the crew in his reassuringly calm voice after a glitchy pit stop: "It's all good. Shake it off." Elliott exited pit road in tenth place.

"Hell, yeah, that's what I'm talking about!" Elliott's spotter Brett Griffin whooped as Elliott surged forward. Twenty laps later, the No. 19 was running sixth. When their car was up front last year, there was a sort of electricity in the pit box, as if it was a surprise. This year, they expect to be up front. Elliott stalled the car coming out of a pit stop on lap 71, losing fifteen positions, but twenty laps later, he had driven the car back to sixth place.

And then, on lap 110, the bad luck that had plagued the team all season made its appearance. I had barely turned around to watch the replay of Elliott's right-front tire blowing out on the track's video screen when a text message from my best friend Vicki (watching at home with fingers crossed) appeared on my cell phone.

"Well, I guess there goes your happy ending. . . ."

It took the team thirty-three laps to fix the damage, but they got the car back on the track and finished thirty-ninth. I might not have a happy ending, but the changes at GEM have given Elliott and his crew a very promising new beginning.

Postscript: In July 2008, Steve Peterson, the technical director of the NASCAR R&D Center, passed away unexpectedly at the age of fifty-eight. Steve's impact on NASCAR is well documented, but his influence extends well beyond the upper echelons of racing. Steve worked closely with a lot of young drivers and engineers, mentoring them and insisting that they make their own safety the same priority he's made it in NASCAR. He had the same passion and concern whether he was talking to teenage go-kart racers or NASCAR drivers. His contributions will be missed.

Notes

Page

2   **atop which ROUSH is spelled out:** After Roush Racing became Roush Fenway Racing, the letters were replaced by the Roush Fenway logo.

4   **at the fall 2006 Bristol race:** Livingstone, Seth. "Bristol Suited to Kenseth's Style." *USA Today*, August 25, 2006.

6   **retains its shape when pushed or pulled:** Technically, ductility refers to how much you can pull on a material before it breaks, and malleability refers to how much you can push on it before it breaks, but we often use ductility to describe resistance to both pulling and pushing.

8   **this process, called "slip":** Every crystal has mistakes, like missing atoms or misaligned planes, which are called defects. Slip is the result of defect motion. A dislocation, for example, is a type of defect in which part of a plane of atoms is missing. Slip happens when a dislocation moves through the crystal, allowing one set of bonds at a time to break and re-form.

25   **organize themselves for maximum ductility:** Aluminum nitride helps keep the steel grains small and produces texture, which means that more of the grains are aligned in the same direction, rather than being randomly oriented. Both features facilitate drawing or rolling of the finished product.

26   **before it even got out on the track:** Newton, David. "'Father' of Car of Tomorrow Confident in His Baby." ESPN.com, Jan 2, 2007, http://proxy.espn.go.com/rpm/columns/story?seriesId=2&columnist=newton_david&id=2717676.

29   **a molded Kevlar-reinforced composite:** Kevlar is a registered trademark of E. I. DuPont de Nemours and Company.

31   **and bond tightly together:** The phenyl groups in Kevlar are so large that they can orient in only one configuration without getting in each other's way. This regularity enables the chains to assemble in an orderly manner. The strength of the material is enhanced because the chains are held together by hydrogen bonding, which is sharing electrons between the hydrogen of one

chain and the oxygen of an adjacent chain. Other types of polymers use a weaker type of bonding called van der Waals bonding and are not as strong.

37 **has a polymer binder and pigment suspended in water:** Most "latex" paints don't actually contain latex. They use tougher (and cheaper) acrylic or polyvinyl polymers. Another chemical called a surfactant has to be added because the polymers generally don't like water. A surfactant is a molecule with one end that likes water and another that likes the polymers, and it serves as a buffer between the polymers and the water.

39 **Cromax Pro paint:** Cromax Pro is a registered trademark of E.I. DuPont de Nemours and Company.

40 **about $7 million worth of thirty-second commercials:** Gage, Jack. "NAS-CAR Races for Sponsors." http://www.wired.com/cars/energy/news/2006/05/70932.

41 **light bounces back at an angle exactly opposite the angle it came in at:** This is another way of saying that the angle of incidence (the angle the light makes with respect to a line perpendicular to the surface) is equal to the angle of reflection (the angle the light makes with the same perpendicular as it's leaving the surface). If the light comes in from the left at an angle of 50° with respect to the perpendicular, it bounces off to the right at the same angle of 50° with respect to the perpendicular.

42 **making that approach a bit expensive:** Reisch, Marc S. "Rainbow in a Can." *C&E News* 2003, 25–27.

45 **special paint called ChromaLusion:** ChromaLusion is a trademark of E.I. Du-Pont de Nemours and Company. ChromaFlair, the pigment used in this and other color-shifting paints, was invented by JDS Uniphase Flex Products Group.

45 **which takes interference one step further:** The flakes are actually five layers: The magnesium fluoride and the chromium metal are deposited on both sides of the aluminum flake; however, light doesn't pass through the aluminum, so there are effectively three layers.

46 **of Motorsports Designs suggested decals:** LeMasters, Jr., Ron. "Teams Trade Paint, Brushes for Vinyl Wraps." February 1, 2007. http://www.nascar.com/2006/news/business/12/19/car.wraps.one.

47 **idea of "wrapping" a car was born:** Lemasters, Jr., Ron. "Racewraps: Devil Is Definitely in the Detailing." February 1, 2007. http://www.nascar.com/2006/news/business/12/26/car.wraps.2.

57 **go off and solidify by themselves:** Gray and white cast iron have cementite or graphite grains embedded in a matrix, like chocolate chips embedded in cookie dough. The matrix can be different phases of iron, pearlite (the layered iron/cementite structure in mild steel), or other iron-carbon phases. The important issue for engine blocks is what form the carbon takes.

58 **extremely narrow range of magnesium concentration:** Magnesium atoms change the way the graphite flakes form. Graphite is *an*isotropic, meaning that

it behaves differently in different directions. The carbon atoms are more strongly connected within planes than they are between planes, which is why graphite prefers forming flakes. The magnesium atoms interfere with the planar growth and the carbon thus forms rounder structures that don't concentrate stress as much as flakes.

59   **only twelve hundredths of an inch thick:** Woodruff, David C. "Why Compacted Graphitic Iron?" *Competitive Production* 2 (2007). http://www.competitiveproduction.com/features/default.aspx?article_id=1388&volume_no=3&issue_no=2.

59   **tens of thousandths of an inch or less are possible:** Cramblitt, Bob. "Reverse Engineering Fine-Tunes NASCAR Engine." *Machine Design* (2004). http://www.machinedesign.com/ASP/viewSelectedArticle.asp?strArticleId=56597&strSite=MDSite.

62   **Mass is a measure of inertia:** Your weight is your mass times the acceleration due to gravity. In the English system, weight is in pounds and mass is in slugs. In the metric system, weight is in newtons and mass is in kilograms.

64   **—meaning higher temperature—is needed:** The water's temperature is a measure of the *average* energy of all the water molecules. The water molecules have a distribution of energies: Some have more energy than average and others have less. When you raise the temperature of the water, you raise the average energy of the liquid, which means that the number of molecules with enough energy to escape the liquid increases because the entire distribution shifts.

65   **bumps rubbing across each other:** Friction is an extremely complex phenomenon that scientists are still trying to figure out. For example, the frictional force is independent of the area of the two objects rubbing together. (Wider tires don't give you more frictional force.) One theory is that only the area of the bumps and dips that actually make contact counts, but not everyone agrees with that explanation. See "New Theory Exposes Cracks in Laws of Friction," by Kristin Leutwyler in *Scientific American,* September 2001, or Jacqueline Krim's article "Friction at the Atomic Scale," in *Scientific American,* October 1996.

66   **fundamental phases of matter:** Plasma is a form of matter in which the positively charged nuclei and the negatively charged electrons separate from each other. Plasmas are found inside neon signs and on the surface of the sun. Some people place plasmas on equal par with solids, liquids, and gases. Other people like to stick with the original "Big Three."

73   **rotate the crankshaft:** The perpendicular distance from the center of the crankshaft to the connecting rod (the piece connecting the crankshaft and the piston) times the force is the torque. Only the perpendicular distance between the force and the center of the crankshaft counts, and that distance changes as the crankshaft rotates.

73 **dividing by 5,252:** There are 33,000 foot-pounds per minute in one horsepower. 5,252 is equal to 33,000 divided by 2π, which is the number of radians in a circle.

75 **in the table:** The values shown in the table for individual cylinder displacement are rounded off. You have to use the un-rounded values when you multiply by the number of cylinders to get the total engine displacement. The values in the left-hand column are from "2007 Toyota Camry Le V6—Technical Specs," Autosite.com.

76 **can both burn a hundred calories of energy:** Everywhere except the United States, food energy is measured in the metric units of joules (pronounced "jewels"). The calorie on food labels is actually a kilocalorie or Calorie with a capital C. I refer to it as a food calorie to avoid confusion.

77 **which is called the stroke:** The volume of one cylinder is $V = \frac{\pi B^2 h}{4}$, where B is the cylinder bore and h is the stroke.

78 **(when the piston is at its highest position):** The compression ratio is different from the displacement because the displacement doesn't take the space in the cylinder head into account.

79 **and 13 percent n-heptane by volume:** This doesn't mean that the 87-octane gasoline you buy has only iso-octane and heptane molecules. It means that whatever mix of gasoline molecules *is* in the fuel has the same overall resistance to ignition as a mixture of 13 percent n-heptane and 87 percent iso-octane.

80 **get 112-octane gasoline:** Technically, there is no such thing as an octane number above 100. If you're talking to a purist, you should refer to "a gasoline with a performance number of 110."

80 **energy the molecule will release:** The energy is determined by the oxidation state of the carbon atoms in the hydrocarbon. Oxidation is when an atom loses electrons (or electron density), as carbon does when it joins with oxygen to make carbon dioxide. The lower the oxidation state of the carbon in the molecule, the more energy is released during combustion when the carbon is oxidized.

87 **"an intoxicated orangutan":** "Pepsi 400—Chevrolet Qualifying Quotes." GM Racing. http://www.race2win.net/wc/02/race/dis2/cqq.html.

87 **landing forty rows into the grandstands":** Clark, Cammy. "Harrowing Wreck Start of Plate Slowdown." *Miami Herald*, October 2, 2005.

87 **rotate 2,160 times per minute:** If you're going 180 miles per hour, you travel 19,008 inches in one minute. Using an average tire circumference of 88 inches, this means the tire makes 19008/88 = 2,160 revolutions in one minute.

96 **circular end facing the wind:** Katz, Joseph. *Race Car Aerodynamics: Designing for Speed.* Cambridge, MA: R. Bentley, 1995.

98 **it is not correct:** NASA has an excellent review of some common theories of lift, with explanations of why many are incorrect, at http://www.grc.nasa.gov/WWW/K-12/airplane/lift1.html.

99   **lift coefficient around -0.42:** "Ford Fusion Specs." Internet Brands, http://
        www.carsdirect.com/research/specs?cat=7&acode=USB60FOC201A0.; Katz,
        Joseph. *Race Car Aerodynamics: Designing for Speed.* Cambridge, MA: R. Bent-
        ley, 1995.

103   **two carbon-fiber roof flaps:** Nelson, Gary, Jack Roush, Gary Eaker, and Stan
        Wallis. "The Development and Manufacture of a Roof-Mounted, Aero Flap
        System for Race Car Applications." Paper 942522 presented at the SAE Mo-
        torsports Engineering Conference, Indianapolis, IN, 1994.

104   **and Kyle Busch at Talladega in 2007:** The latter two events (Stewart and
        Busch) occurred during NASCAR Nationwide Series races.

105   **keep up with a much faster one":** Higgins, Tom, and Steve Waid. *Junior
        Johnson: Brave in Life.* Phoenix, AZ: David Bull, 1999.

106   **sucked that back glass right out":** Higgins, Tom. "Gone! With the Wind."
        http://blogs.thatsracin.com/scuffs/2006/01/gone_with_the_w.html. Jan 25, 2006.

106   **in each cell are solved numerically:** Some equations are so complex that you
        can't solve them algebraically. You have to use numerical techniques, which
        often yield approximate solutions.

110   **he'll *want* you to pass him":** LeMasters, Jr., Ron. "Working the Draft."
        Stock Car Racing (1999). http://www.stockcarracing.com/thehistoryof/11298
        _working_draft/.

111   **It's just going to slide":** Boone, Jerry F. "Bad Air." http://www.stockcarracing
        .com/techarticles/general/134_0212_racing_aerodynamics.

111   **to overcome that drag:** Eight is 2^3. If you tripled your speed, you'd need 3^3, or
        twenty-seven times more power.

115   **splitter is made from Tegris:** Tegris is a trademark of Milliken & Com-
        pany.

116   **the drag closer to that of the old car:** Carney, Dan. "Car of Tomorrow's Day
        Arrives." *AEI*, May 2007, 61–63.

117   **concentrated almost entirely near the splitter:** Grissom, Glen. "Wind Tun-
        nel in a Can," *Max Chevy Magazine* 36, no. 3 (2007): 36–38.

128   **temporary blindness and blackouts:** Lorditch, Emilie. "Race Car Drivers
        Dizzy Over Physics." *Inside Science*, College Park, MD: American Institute of
        Physics, 2001. http://www.aip.org/ISPS/reports/2001/015.html.

128   **Firestone Firehawk 600 CART:** CART stands for Championship Auto Rac-
        ing Teams, which is an open-wheel series now known as the Champ Car
        World Series.

128   **straightaways before turning again:** Guedry, Frederick E., Anil K. Raj, and
        Thomas B. Cowin. "Disorientation, Dizziness and Postural Imbalance in Race
        Car Drivers, a Problem in G-Tolerance, Spatial Orientation or Both." Pensa-
        cola: University of West Florida Pensacola Institute for Human and Machine
        Cognition, 2003.

129   **how much turning force:** What I'm calling turning force is usually referred to

as *lateral* force because it is perpendicular to the direction in which the car is moving. When a car is cornering, the turning force is directed toward the center of the turn.

132   **(CG for short):** The center of mass and the center of gravity are in the same place if you are in a uniform gravity field. In astrophysics (for example), this is not always the case, and there is a difference between center of mass and center of gravity. NASCAR races take place in uniform gravity fields, so we'll treat them as the same thing and use CG, because that's what the race car engineers usually use.

133   **weight shifts from front to back and from left to right:** The forces that stop or turn the car act at the tires. Those forces create torques about the car's center of gravity and those torques are responsible for the load transfer. "Why Does an Accelerating Car Tilt Upward?" Everett Hafner, *The Physics Teacher* 16 (2), 122 (1978), gives a concise explanation of the physics of load transfer.

135   **Josh Browne, team director:** Gillett Evernham Motorsports has a slightly different organizational structure than other teams. Most teams call the person in Josh's job the crew chief.

140   **110 to 125 dB inside:** Van Campen, L.E., T. Morata, C.A. Kardous, K. Gwin, K.M. Wallingford, J. Dallaire, and F.J. Alvarez. "Ototoxic Occupational Exposures for a Stock Car Racing Team: I. Noise Surveys," *Journal of Occupational and Environmental Hygiene* 2, no. 8 (2005): 383–90.

155   **hits the inside of the tire, it exerts a tiny force:** In addition to air molecules pushing out from the inside of the tire, air molecules outside the tire are pushing in. The tire pressure you measure is the difference between these two opposing pressures.

165   **spring is compressed doubles as well:** Everything I say about compression also applies to stretching a spring, but springs don't get stretched in the car suspension unless something has gone wrong.

175   **sixteen different configurations:** The total number of configurations is 2^4. There are two independent choices for each wheel, and there are four wheels. If he had four choices per wheel, there would be 4^4 possible combinations, which is 256.

193   **The mass of the car and driver remain constant:** If a part—like a bumper cover—comes off the car, its mass changes; however, the change in mass is small in most cases. To first order, the mass is constant and we can consider only the change in velocity.

194   **Vectran:** Vectran is a registered trademark of Kuraray Co., Ltd., Tokyo, Japan.

194   **developed by Celanese Acetate:** Kuraray Co., Ltd., bought Vectran from Celanese Advanced Materials in 2005.

196   **and an inner Nomex lining:** Nomex is a registered trademark of E. I. DuPont de Nemours and Company.

200   **describing a snowflake by its diameter:** The analogy that crashes are like
       snowflakes in their individuality was suggested to me by the writer Allen St.
       John.

201   **than the car on the right:** The angle is measured with respect to the wall. A
       90-degree angle would be head-on.

210   **before selecting IMPAXX:** IMPAXX is a registered trademark of The Dow
       Chemical Company.

210   **The best-known polystyrene foam is STYROFOAM Brand Foam:** STYRO-
       FOAM Brand Foam is a trademark of The Dow Chemical Company or an af-
       filiated company of Dow.

211   **a new hyperelastic foam called FlexAll:** FlexAll is a registered trademark of
       Battelle.

224   **with a lower atomic number:** This is generally true. Some isotopes violate this
       assertion.

225   **car rolls in the turns:** When you move the wheels to the right, more weight is
       located to the left of the roll center, so this causes more weight transfer around
       the corner.

232   **rash or careless behaviour":** Walker, Scott, Timothy R. Ackland, and Brian
       Dawson. "The Combined Effect of Heat and Carbon Monoxide on the Perfor-
       mance of Motorsport Athletes," *Comparative Biochemistry and Physiology A*,
       128 (2001): 709–18.; Chesson, B.J. "Evaluation of Personal Cooling Suits."
       Thesis, Western Australian Institute of Technology, 1984.

232   **you sort of accept":** Goad, Libe. "Don't Call Him Ricky Bobby." In *Video
       Game Features*: AOL Games, 2006.

234   **CarbonX, which was developed by Chapman Innovations:** CarbonX is a
       registered trademark of Chapman Innovations.

237   **average car exhaust is CO:** Poels, E.K., and D.S. Brands. "Catalytic Purifica-
       tion of Exhaust Gases in Internal Combustion Engines." http://www.science
       .uva.nl/~gadi/pdf_files/Exhaust_gases_purification.pdf.

238   **I realized what was happening":** Hinton, Ed. "Strength in Numbers." *Sports
       Illustrated* 2, no. 16 (1997): 86–87. http://sportsillustrated.cnn.com/features/
       1997/nascar/nshend.html

238   **scienced out to a T yet:** Mulhern, Mike. *Winston-Salem Journal,* March 31,
       2007.

238   **to decrease motor skills:** Spencer, Lee. "The Gas NASCAR Didn't See."
       Sporting News, February 3, 2003, p. 53.

240   **that worked at lower temperatures:** "NASA Develops Catalyst for NAS-
       CAR." *FLC Newslink*, February 2006.

250   **polymer called LEXAN:** LEXAN is a registered trademark of GE Plastics.

250   **Daniel Fox in 1953:** Dr. Hermann Schnell of Bayer (Germany) applied for a
       U.S. patent on a very similar molecule the same year GE filed for the patent on
       LEXAN. Fox generally is credited with the discovery, but it is not unusual for

two different people working in two different places to come up with similar—or even identical—discoveries.

251   **can last four to six months:** Pappalardo, Joe. "NASCAR Windshield Laminates Gaining Military Following." *National Defense*, 2004.

251   **DuPont polymer called Mylar:** Mylar is a registered trademark of E. I. DuPont de Nemours and Company.

252   **air gap between the layers:** To be precise, 96 percent of the light goes through the tear-off. You lose 4 percent of that 96 percent when the light reflects coming out of the film, so you actually lose 7.8 percent of the light.

269   **"I went from being stubborn":** Smith, Marty. "Sadler swallows pride, realizes this old dog needs to learn new tricks." http://sports.espn.go.com/rpm/nascar/cup/columns/story?columnist=smith_marty&page=DoorToDoor20080730.

Acknowledgments

Many, many thanks to all of the people who were generous enough to spend time answering all my questions. I've mentioned most by name in the book and won't repeat them here only for reasons of space. A number of people went way beyond the call of duty. Extra special thanks to Josh Browne, Gary Eaker, Mike Fisher, Chad Johnston, Andy Randolph, Dean Sicking, Mark Thoreson, Eric Warren, and Nick Woodward, who spent extra time verifying details with me. Thanks also to Tom Gideon, Bob Hoekstra, Bob Hubbard, Jim Kontje, and Todd Meredith, who provided important background information.

Many thanks to the people at NASCAR: Heather Greene, John Farrell, Jennifer White, and Andrew Giangola, who helped me immensely with logistics. Andy Moffat and Heather Petry from Gillett Evernham Motorsports; Heath White, Spencer Andrews, and Denny Darnell from Clear!Blue; Jesse Essex and Megan Whiteside from Hendrick Motorsports; and Jon Edwards from PPR Plus also provided valuable assistance arranging visits and interviews.

Special thanks to Josh, Chad, Kirk, Ramon, Chris, Tom, Allen, Swifty, Kiwi, and Tony for letting me be underfoot, answering my unending stream of questions, and making sure I didn't get run over in the garage. Wherever you go, whomever you race for, I will be cheering for you. Thanks also to the guys on the Nos. 1, 16, 20, 21,

43, and 45 teams who were next to the No. 19 car at the races I attended. Your patience and good humor was much appreciated.

It is surprising how long the journey from idea to paper is. This particular odyssey began with my longtime friend Jack Hehn, who asks me to be on a lot of committees and come to a lot of meetings. One reward for agreeing has been meeting interesting people like Jennifer Ouellette who, in addition to my admiration for her own writing, has my eternal thanks for telling me to "just do it." Jennifer introduced me to my agent, Amanda Mecke, who has been a wonderful teacher and supportive friend throughout the entire process. My editor at Dutton, Stephen Morrow, deserves special thanks for dealing with my first-time-author questions and worries. My copy editor, Gary Perkinson, has my great appreciation for making me look much more literate (and grammatically correct) than I actually am.

I cannot thank Vicki and Mark Plano Clark enough for their friendship, advice, support, and pet-sitting. Their patience with my finding a way to insert something NASCAR into every conversation for the last fifteen months (and counting) is monumental and much appreciated.

My greatest debt of gratitude is to my husband, Bob Hilborn, who is—and has always been—unconditionally supportive. From freezing his toes off at the Las Vegas Motor Speedway to debating the intricacies of carbon-dioxide lasers, Bob is a constant partner and friend. His faith in me allows me to accomplish things I would never have dared attempt by myself. I count myself blessed every day because he is part of my life.

Index

Note: Page numbers in *italics* denote charts and illustrations.